# PREPPER'S TOTAL GRID FAILURE HANDBOOK

## ALTERNATIVE POWER, ENERGY STORAGE, LOW-VOLTAGE APPLIANCES, AND OTHER LIFESAVING STRATEGIES FOR SELF-SUFFICIENT LIVING

### ALAN AND ARLENE FIEBIG

Ulysses Press

Published in the U.S. by
ULYSSES PRESS
P.O. Box 3440
Berkeley, CA 94703
www.ulyssespress.com

ISBN: 978-1-61243-637-1
Library of Congress Control Number 2016950679

Printed in Canada by Marquis Book Printing
10 9 8 7 6 5 4 3 2 1

Acquisitions Editor: Casie Vogel
Managing Editor: Claire Chun
Project Editor: Caety Klingman
Editor: Shayna Keyles
Proofreader: Lauren Harrison
Index: Sayre Van Young
Front cover/interior design and layout: what!design @ whatweb.com
Cover artwork: solar panels © Wouter Tolenaars/shutterstock.com; background grain © Nerijus Juras/shutterstock.com; gas generator © yevgeniy11/shutterstock.com; pedal-powered generator © Kenneth Torino; flooded lead acid battery © U.S. Battery
Interior photos: page 17 © Bill Fehr/shutterstock.com; page 27 © Wikimedia; page 31 © grmarc/shutterstock.com; page 150 © James Steidl/shutterstock.com

Distributed by Publishers Group West

*Dedicated to our Lord and Savior, Jesus Christ, who counsels us to prepare for dark days ahead:*

"It was by faith that Noah heard God's warnings about things he could not yet see. He obeyed God and built a large boat to save his family. By his faith, Noah showed that the world was wrong, and he became one of those who are made right with God through faith."

*—Hebrews 11:7*

# CONTENTS

# PROLOGUE: A CAUTIONARY TALE

Drew slammed the door of his car a bit harder than necessary, rattling the windows. This was not what he had expected as an assignment when he started with the *Boulder Prodigy* newspaper. If it weren't for the crazy tinfoil hat–wearing readers and the NOAA/NWS Space Weather Prediction Center (SWPC) being located in the same town as the paper, he probably would not have been saddled with this wild goose chase. Being low in the paper's pecking order probably didn't help. To top things off, his car's GPS system hadn't been able to get a single satellite on lock for two days now, so he had been driving in circles for the past half hour just to find the parking lot. But he finally arrived and headed in to conduct a worthless interview where he would try to answer all the questions that had been phoned in to the paper over the past couple nights.

Thursday night had arrived with extremely bright, shimmering, moving curtains of light. Most people in Colorado were familiar with the aurora borealis, the Northern Lights. But no one had ever seen anything like these lights, which were bright enough to read by. The following morning, the newspaper's voicemail box was filled with frightened messages from readers asking if it was the end of the world. Drew's editor felt that they could sell some papers by making an effort to answer their readers' questions.

"Come on, people, what ever happened to common sense?" Drew thought. "It's just lights, right?"

Drew made an appointment with NOAA's public relations person and resigned himself to what he expected would be a day of listening to some science-babble about lights in the sky. As he entered the lobby, Drew was drawn to the gallery where there were poster-sized photographs taken by the Hubble telescope. It was too bad they didn't have a gift shop, but the SWPC wasn't exactly set up to cater to the public. Drew thought that they took their security rather seriously for a bunch of squints. After passing the checks on his press credentials, he was led into the office area and introduced to Ms. Kelly Sinclair.

"Please," she said, coming from behind her desk, "just call me Kelly. I try to work closely with as many members of the press as I can so the public can be aware of what we do here."

"Well, Kelly, what exactly is it that you do here?"

"I know this is going to sound like a canned response, and that's because it is. We continually monitor and forecast, as best as we can, the space around the Earth and anything affecting it. Much of what affects the space around our planet originates from the sun, which I think is why you are here today. SWPC is the official source of space weather alerts and warnings for the United States."

Drew was confused. "How can the sun be responsible for what has been driving the panic of the last two nights? It was dark out then; the sun had set by the time the craziness started."

"Well, Drew," Kelly explained, "two days ago we observed a massive solar flare on the surface of the sun, and unfortunately it was aimed straight at the Earth. Solar flares are our solar system's largest explosive events, releasing large amounts of radiation that travel from the sun to the Earth in about a half hour. At the time of this flare, North America was facing the sun. If a flare is traveling toward the Earth, it will hit whichever part of the Earth is facing the sun at that time. You may not have noticed anything from the impact, but a lot of our scientific equipment in orbit sure did. We lost over a third of our GPS satellites, just to give an example."

Drew suddenly started paying attention. "So is that why my car's navigation system isn't working right?"

Kelly nodded. "Exactly! But I haven't answered your question about the strange lights yet." She paused to gather her thoughts, not wanting to lose Drew's attention; she noticed that his heart wasn't really in the interview. "Solar flares are often accompanied by coronal mass ejections, or CMEs, which move much slower than the flares that cause them. CMEs are made up of a bunch of sun junk, like large clouds of plasma, X-rays, and gamma rays. A CME can take one to five days to arrive at the Earth, where it has a large effect on the Earth's magnetic field and atmosphere. The energy of the CME can cause particles in the atmosphere to glow, resulting in those lights everyone has been seeing—the Northern Lights."

"Okay, thanks. I'll let our readers know that this CME is nothing to worry about and that it just causes pretty lights. It's not the end of the world. I guess we're done! Thanks for your time."

Kelly shook her head. "Whoa, not so fast. I never said they are harmless. Why do you think this organization exists here? Why do you think we issue alerts?" Drew sat back down, a look of concern on his face. "As I said," Kelly continued, "a CME arrives with a lot of energy, which can affect our magnetic field and the atmosphere below it. That same energy can also reach the ground, and it can dump energy into any long metallic structures like pipelines, power lines, and phone lines, for example."

Drew started taking notes again as Kelly continued. "Just like hurricanes and tornadoes, solar flares and geomagnetic storms around our planet have ratings to indicate how severe they are. Flares come in one of three major sizes: C-Class, M-Class, and X-Class. C-Class flares are very minor and not noticeable to most people. M-Class flares are mild and cause some interference with communications. X-Class flares are severe, and they can cause electrical damage and power outages. Within each of the three classes, each flare is also given a rating between 1 and 9, with 9 being the most severe.

"The same goes for geomagnetic storms, which are caused by a CME hitting the Earth. They are rated from a minor G-1 to a severe G-5. Usually, the more severe the solar flare that produces the CME, the more severe the geomagnetic storm will be when the CME hits the Earth."

Kelly paused for Drew to catch up with his notes and then shifted direction. "Back in 1859, the Earth was hit with a massive CME that caused lots of damage. The United States didn't have advanced technology back then, but our early telegraph system had strung miles and miles of wire up in the air, running right alongside miles and miles of iron railroad."

"And?" asked Drew.

"The planet experienced what is known as the Carrington Event, named after the astronomer Richard Carrington, who observed the solar flare that accompanied the CME. Another CME had hit the Earth a day or so earlier, weakening our protective magnetic fields. The Carrington CME hit the Earth so hard that the Northern Lights woke up miners in the Rocky Mountains, who started preparing breakfast thinking it was dawn." Kelly grew more serious as she continued. "As I said, we didn't have any computers, satellites, Internet, or even power grids back then; just the telegraph. And when the CME hit, telegraph operators were physically shocked, and fires started along the lines. Just imagine what would happen in today's society." She paused for a moment. "Bank ATMs—gone. The power grid—gone. Cars might not start, but even if they did, gas stations couldn't pump gas without power. Cell phones, food delivery, refrigeration—all gone. No water would come from the faucets because the water pumps would no longer run. A CME like that just missed the Earth on April 28, 2014. How long do you think we will be so fortunate?"

Drew sat silently for a minute. "Okay, how long would the power be out? Once the CME fades, it all comes back, right?"

"Nope, not a bit of it. The electronics would all be fried and would have to be replaced. Ever had lightning strike near your house?"

"Sure, a couple of times. And I lost my television and my computer, had smoke coming out of them. And you know what they say about electronics, they are magic devices that run on smoke, so never let the smoke out!" he laughed nervously.

Kelly was really trying to get through to him with just how serious the magnitude of a CME could be. "Okay, let's say that after the lightning struck, the transformer on the pole outside your house that connects you to the power grid got damaged. I bet the power company sent out a truck to replace it, didn't they?" Drew nodded. "Well, what happens when every transformer in the country is damaged? There is no gigantic stockpile of transformers. And the trucks that do the work will eventually run out of fuel. The factories that make the transformers have no power to run their equipment. Deliveries of the parts and materials needed to make more can't be shipped. The power will be off for a long, long time. Eventually, we'll start to run out of food. Farmers might have enough food for themselves, but they won't be able to get it to the stores. And without fuel, plantings and harvests will be small."

Drew had grown pale. "You are not painting a pretty picture here; this is a doomsday scenario you are describing."

"Exactly. What was a minor inconvenience back in the 1800s is now the end of the world as we know it." Tapping her pen on the table for emphasis, she continued. "The 'as we know it' part is the problem. Back in the 1800s, most people lived either on or near a farm. Cities and villages were small enough and close enough together for people to get what they needed from someone nearby. Markets got their food locally, and bartering often made the need for money unnecessary. Certainly, credit and debit cards were not needed. That is not how we currently know our world, and if the technology and modern devices we now depend on suddenly become scrap, we will be in a world of hurt."

Drew, looking at the tablet he was taking notes on like it might suddenly melt, put it away and stood. "Well, I thought there wouldn't be a story here, but now I think I may make tomorrow's front page. Thanks for your time, and I sure hope what you described doesn't happen in my lifetime."

Kelly shook Drew's hand and said, "Don't get too comfortable. Yesterday, we saw a second flare erupt from the same part of the sun. We don't know yet if there was a second coronal mass ejection; it's too early to tell."

On his way home, Drew decided that to be on the safe side, he would stop at the nearest home improvement store and pick up a small generator, some gas cans, and maybe some other alternative power devices like oil lamps. "It never hurts to be prepared," he thought; and besides, the winter storms near the Rocky Mountains can often knock out the power. "Maybe some freeze-dried camping food and some water jugs would be a smart purchase, too. Perhaps I need some tin foil to make a hat for myself. These 'preppers' are looking a bit smarter than I used to think they were. I never questioned the status quo before. I flip the light switches, I start the car, I fill up the gas tank, I go to the grocery store, always just taking them for granted."

Late that night, Drew was woken by shouting outside of his window, coming from the street below. He got up and looked out his window but could see nothing in the complete darkness. He felt for the bedroom light switch and flipped it on, expecting the lights to come on as they always had in the past. He sank to the floor, his head in his hands, heavy with understanding. The CME had happened. He so wished that he had been prepared...

# INTRODUCTION

It was a dark and stormy night...

Well, maybe not stormy, but I've always wanted to start a book that way! And it *was* dark...

We had big delays in closing on our off-grid mountain property because of an error in the deed's legal description. Had it not been corrected, we would have wound up with an unknown piece of property miles from the land we were actually trying to buy. After a few more bumps and delays, the actual closing process was quick and painless. With the deed in hand, we drove into the mountains to the property and did not arrive until it was already dark.

We lit the kerosene lamps we brought with us to fill the room with light, and we were then able to set up a folding table, chair, and air mattress for our first night living off-grid! We wanted to share this event, so we created a quick Internet setup and used battery power for the laptop. We successfully made our first off-grid post on our blog site, www.off-grid-geeks.com.

But we knew these power sources would not last. We needed a renewable, self-reliant, alternative energy solution. So the next morning, we set up a 15-watt solar panel in the cabin's only south-facing window and connected it to a very simple charge controller and a small battery. It wasn't much, but we now had a means to produce power—power that would replenish itself. That's how it all started for us: with one solar panel. Over the next four years, we learned a lot about self-sustaining power and built our own off-grid system. This book will explain how we were able to forgo commercial grid power and instead create a self-reliant, renewable alternative in a remote location in the Ozark Mountains.

# WHAT IS OFF-GRID?

An off-grid home is not connected to any public utilities. All the modern conveniences of society that are generally supplied by a municipality are either deemed unnecessary by the homeowner or are otherwise provided without connecting to an outside source. The term "off-grid" is derived from the fact that you are not connected to the electrical power grid in your community, though an off-grid life can also mean being disconnected from all other public utilities including water, sewage, natural gas, and landline telephone. Any conduit that brings one of these public services onto your property connects you into the grid; to be fully off-grid, you must break all these connections.

In our case, we chose to break virtual grid connections as well as physical ones. We disconnected with services such as trash collection, package delivery, newspaper delivery, and mail service. Going off-grid is different for everyone, and can be done in baby steps or by going cold turkey, which is what we did.

# WHY GO OFF-GRID?

We chose to make the jump from a fully connected house to a house with no existing grid ties, meaning we had no electricity and no running water. However, we never planned to live without those modern conveniences. Instead, our goal was to continue our high-tech lifestyle by taking care of all our own needs.

So why did we chose to go through all the hard, time-consuming work of setting up a self-sufficient home in the middle of the woods? We had several reasons.

- If the grid eventually goes down due to natural or manmade causes, we will still have all our daily modern conveniences.

- We wanted to undertake the adventure of setting up and living a self-reliant lifestyle.

- The more ways you move off the grid, the more anonymous your life becomes.

A fully off-grid lifestyle is not for everyone. Many of our friends who are preppers do not want to live off-grid unless they have to. They set up solar electric systems similar to ours while remaining tied in with the public electric company. In this manner, their alternative energy is ready when it is needed and they can use it in certain areas of the house to become familiar with it before "crunch time."

# PREPPING AND ALTERNATIVE ENERGY

Whether you are new to prepping or have always lived a lifestyle where preparing for emergencies and possible disasters was a part of your day-to-day life, you probably soon realized just how many different types of prepping you can do. There are seemingly endless avenues you can travel down when it comes to prepping, and you need to prioritize which are most important for you and your family.

One way to decide on your prepping priorities is to consider what will be less available in the event of a cataclysmic disaster and if you will be able, or willing, to live without those things. Thinking this way will give you a list of necessities, and you can then start prioritizing by affordability. Due to cost, many preppers move alternative energy from their needs lists to their wants lists, but in this book, we will show you how having your own sources of electricity can actually be very affordable.

People commonly assume that their power consumption during a grid-down situation will be the same as it is when they are on the grid, though actually, it can be much less. Think about an extended grid-down situation—you probably won't need a very large refrigerator because you won't have much to put in it. After all, it's not like you'll be driving down to the grocery store to fill it up on a regular basis. Just as you sat down and created your necessities list, it's a good idea to create a list of what you will realistically be using your alternative energy for.

For example, even though we still run our high-tech businesses from our off-grid house, our power consumption is much lower than when we were grid-tied because of certain choices we made. We have no dishwasher, no washer/dryer, and no air conditioning. Our businesses are necessary for us to make a living, but these luxuries are not. It's easy for us to hand wash our own dishes and clothes (page 44) and to cool our home using natural means. High-efficiency options for lighting and appliances also help (see Chapter 14).

Preppers have many different motivations that drive their preparations. If you look at all the various triggers that could bring down modern society, you'll see that the electrical grid would be a victim in all potential situations. Experts and even government officials regularly tell us that our electrical grid is extremely fragile and vulnerable to many types of disasters and attacks. Our goal in this book is to provide you with the information you need to be able to prepare for the day the country's lights go out.

## WHAT DO SELF-RELIANCE, HOMESTEADING, AND PREPPING HAVE TO DO WITH EACH OTHER?

Arlene loves math and often views life as one big Venn diagram. A Venn diagram is a way to visualize how sets of data overlap and interact. Take a look at this example of how self-reliance, homesteading, and prepping overlap.

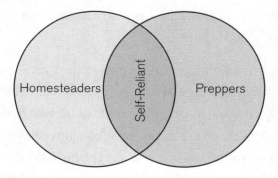

There are homesteaders, there are preppers, and the overlap is self-reliant people.

Not all homesteaders are self-reliant. They may focus on growing their own organic food supply, decreasing their carbon footprint by living green, or maybe they just enjoy farming. To do these things, they run their equipment on gasoline, rely on the local electric company for their electricity, and buy their fertilizer and feed at the store.

Similarly, not all preppers are self-reliant. Preppers often focus on storing up enough supplies to make it through whatever hardship they believe to be coming, such as a storm, war, terrorist attack, or economic collapse. But some preppers rely solely on their stores instead of producing any of their own supplies.

Some homesteaders and preppers are self-reliant, and this is where these two groups overlap. Their shared goal is to be able to provide for themselves in a renewable manner, without overly relying on outside sources. The self-reliant strive to grow their own food by using heirloom plants whose seeds can be saved and replanted. They have a way to supply their own water, perhaps from a well, a stream, or through rainwater collection. They provide an off-grid means of staying warm in the cold months, or they choose to live where it doesn't get cold. If they have livestock, they grow their own feed. Most important, they find renewable solutions to their power needs and do not rely on outside sources of fuel such as gasoline, diesel, or propane.

We want to teach methods of alternative and renewable electrical power to those wanting to be self-reliant, whether they are homesteaders, preppers, or both. This book will teach you about the options available for generating electricity when the grid goes down, how to leave a smaller carbon footprint, and how to make educated and economical choices as you prepare for a self-reliant, grid-down lifestyle.

# HOW TO READ THIS BOOK

This book is designed to give you a thorough understanding of alternative power systems and teach you how to implement them. The chapters that discuss batteries, battery banks, and charge controllers are applicable to many different sources of alternative energy, but we have found that solar electric power is the best solution in our location. Therefore, the primary focus of this book is on setting up an affordable solar-power system and adjusting your daily power needs to match the amount of power you can generate from your self-reliant system. We do include discussions on both wind and water-based power generating systems so that readers can become familiar with the various physical requirements that need to be considered when contemplating these alternatives. While we, as a husband and wife team, were able to purchase our solar system by piece and build it ourselves, this would not have been possible either with the home wind turbine or micro-hydro generator alternatives we discuss in this book.

We are not armchair authors who have only learned about these things in books or classrooms. We started four years ago by jumping completely off the grid, learning through hands-on experience how to build and perfect the systems that allow us to survive. We feel that this ride-or-die approach had many benefits. For one, it would have been harder to start on the grid and try to wean ourselves off it. There is a saying that the amount of stuff you have will expand to fill the amount of space you have. If you start with a lot of available electricity from the grid, your consumption will likewise grow. This will make it difficult to pare back when switching to alternative power that, while adequate to live on, is not as plentiful as grid power.

We hope to help you bypass some of the beginner steps we took along the way as we learned and refined our system. We have learned from our mistakes, and you can too! With the help of this book, you'll make a few less of them than we did.

# PART ONE

# EARTH, WIND, AND FIRE: GENERATING RENEWABLE ENERGY

# CHAPTER 1
# AGAINST THE CURRENT

## Overview of Alternative Power Systems

There are many ways to create electrical power, also referred to as energy. Of course, we know that power cannot be created; it can only be transformed from one form to another. That's what we mean when we talk about "creating" power. All of the various ways of creating electricity simplify to transforming one form of energy into another. The most common means is through a generator.

To talk about generators, we first need to talk about magnets and coils of wire. In 1820, Hans Christian Oersted discovered that if electricity was run through a coil of wire, a magnet was created. These physical interactions result in electromagnetism. Oersted's discovery eventually led to the invention of the electric generator, which is a device that creates electrical power by transforming mechanical energy into electrical energy. This is often done with a spinning magnet inside a coil of wire.

If an electric generator produces direct current (DC), as the earliest kinds did, it is called a **dynamo**. If an electric generator produces alternating current (AC), then it is called an **alternator**. Alternators and dynamos are both types of **generators**. Generators use fuels like gasoline, diesel, or perhaps propane to run an engine, which spins an alternator to provide you with 120 volts of AC power to use as you choose. We use the term "generator" here to mean a complete device that produces electricity; it includes both the engine and the alternator it spins. (For more on AC and DC power, see Chapter 13.)

This leads us to our main topic—the different forms of mechanical energy that can be converted by dynamos or alternators into electrical power. As most of the systems that follow produce direct current, we will use the term "dynamo" when referring to their generators. The dynamo stays the same, but the method we use to spin it varies. Some of these methods include water, wind, and sunlight: earth, wind, and fire.

# MICRO-HYDRO GENERATORS

Hydroelectric power is electricity created from the movement of water. The motion of the water current along with the mass of the water turns a dynamo, which in turn produces electricity.

It is unlikely you have the resources to build a hydroelectric plant on your property. However, what you might be able to use is what is called a micro-hydro, or a micro-hydroelectric generator. As its name implies, a micro-hydro is a small system that transforms the flow of water into electricity. All you need is a good, solid, consistent flow of water and a micro-hydro generator.

In a micro-hydro generator, the water flow spins turbine blades that turn the dynamo. A quick search on the Internet for micro-hydro generators will provide you with an extensive list of vendors, as well as informational websites.

Along with researching the various types of available micro-hydro generators, you should spend time determining if your location lends itself to this type of alternative energy solution. While you do not need to live near a river or rushing waterfall, you do need to have an adequate water source to provide the force needed to generate power.

This brings us to the topic of **head**, which is the change in elevation between the upstream source of your water and the downstream location of your micro-hydro generator. The greater the head and the greater the supply of water, the more power you can produce. Note that your water source does not have to be flowing, such as with a stream or river. It can be a pond or even a pool that has been made by damming a small stream, just as long as it is always available and is at a high enough elevation above your turbine to create the necessary head. If you have a water source high enough, such as a pond up on a ridge or hill, then gravity can pull that water down through a pipe with enough speed to spin the blades of your turbine.

The water pipe that ferries the flow between the source and the turbine is called a **penstock**. It's usually 2 to 4 inches in diameter. For a micro-hydro to work, you need to have enough head to ensure a fast water flow, as well as enough water volume to fill your penstock and keep it flowing without its running out. If you use a pond, there must be a constant source of water flowing into it; you don't want to just drain the pond and then run out of water. Remember that you need enough waterpower to transform one form of energy (the flow of water) into another form of energy (electrical energy).

As a basic test, use this formula to see if micro-hydro might work for you:

HEAD in feet × FLOW in gallons/minute ÷ 10 = WATTS

Usable micro-hydro systems have been built with as little as 10 feet of head, but when thinking about both flow and head, more is always better. Higher head results in faster-moving water, and faster-moving water means more water is being pumped into your micro-hydro. Both of these factors mean that a larger source of water is required.

While a micro-hydro system is easily operated, it is not easily designed and constructed. For that reason, we recommend that if you feel you have met the basic requirements for head and flow, you should find yourself a commercial vendor of micro-hydro systems to evaluate your site and help you design an appropriate system.

Remember the rule of preparedness that "two is one, and one is none:" always have a backup. Micro-hydro systems have mechanical moving parts that wear, so be sure to keep a supply of extras of anything that might need replacing down the road.

As with all renewable resources, you will have good energy-producing days and bad production days. You may have had a recent flood that fills up the streambed, or you may have a drought. The pond sourcing your water flow may be full, or it may be empty. But overall, micro-hydro generators tend to produce power 24 hours a day, compared to solar's four to eight hours a day. Your generator may produce less power at any given moment, but like the tortoise that beat the hare, it will just keep plodding on day and the night, so keep that in mind when considering this type of generator.

# HOME WIND TURBINES

In theory, the only difference between hydroelectric power and wind turbines is that **turbines** use wind currents rather than water currents. But in practice, it's never that easy...

Wind currents can change in direction, which is something water currents never do unless there is a flood. Wind currents can also change in speed and force much more often than the speed and force of water flow changes. Wind power makes for a good companion to solar energy. Especially in stormy weather when the sun doesn't shine, you'll often get high winds, and on those dog days of summer when the wind is gone, you'll have lots of sun.

Just as micro-hydro requires a good source of flowing water, wind turbines require a substantial source of flowing wind. The long-held dream of using a turbine with a low wind speed has not been realized for one basic but significant reason: You can't create energy; you can only transform it. In other words, you get out what you put in. If you can only put in a weak wind that blows at a low speed, you will only get weak electrical power.

When you take into consideration all the parts of a wind turbine, they add up to a certain amount of mass. It's important to keep mass in mind when trying to transform power into energy. According to Newton's First Law, an object at rest wants to stay at rest, meaning that object will not move unless acted upon by an outside force. Say you have a wind turbine set up in your backyard and the weather is calm. No wind means the wind turbine has stopped turning; it is at rest, and according to Newton's First Law, it wants to stay that way. So, if a nice little summer breeze pops up, say 5 mph, that turbine is just going to keep on resting. A wind turbine requires a certain amount of wind speed just to get it to stop resting and start turning to produce some power. The amount of wind needed for the turbine to start producing power depends on how much it takes to get the turbine to start spinning. The wind needs to be blowing hard enough to overcome the mass of the turbine blades as well as the friction of all the turbine's moving parts.

The speed at which a wind turbine will start producing power is called its **cut-in speed**. When you look at your various options for home wind turbines, the cut-in speed is an important number to look at. To answer the question running through your mind, the most common minimum speed is 10 mph. The U.S. Department of Energy maintains detailed wind maps you might consider that you can find in Resources on page 177.

To better understand how a wind turbine works and to comprehend one of the potential dangers of turbines, we need to take a few minutes to explain electrical load and its impact on the turbine's rotational speed. The electrical load is the amount of power being drawn from the turbine (or any other electricity-generating device). That electricity could be flowing into batteries to charge them, powering tools for a project, running the television while you watch the big game, or a combination of any or all of these activities. As this power is drawn from the turbine, a **load** is placed on the parts that are physically turning. Similar to friction produced by moving parts, the load causes the blade rotation to slow. Under typical loads, the speed variations would probably not be noticeable, but if you have a heavy load to power, the blades may not spin without high wind speeds. Similarly, if the wind slows, your turbine could stop. In these situations, the wind is not strong enough to overcome the turbine's mass, friction, and load to turn the blades effectively.

For a better idea of how load works, think about driving your car down the road. You have your accelerator pressed down just far enough to maintain your speed, but what happens when you start up a hill? You will need to apply more power to climb the hill, so you feed your car's engine more fuel by pressing down harder on the accelerator—in other words, you overcome a larger load by producing more energy. If the hill is steep enough and your car's engine just can't produce enough power—or if the load is too large and you can't make enough energy to meet that load—the car could come to a stop.

We also have to consider the opposite problem: What happens when there's too much wind or too light a load? Let's imagine for a moment a day when you don't need any power. Perhaps you are away from home. Your battery bank is all full, so there are no demands being placed on your turbine; the wind has an easy job of spinning the turbine. Then a storm comes up, the wind builds, and the blades start turning faster and faster. With no load to slow it, the speed can build so high that the turbine burns out, throwing blades and other parts through the air like ninja throwing stars. Some wind turbines have built-in braking, but others do not; they rely on **dummy loads**.

Look back a few paragraphs to where we spoke about mass, friction, and load. We mentioned that a heavy load could slow down your turbine. In this case, a turbine-stopping load is exactly what you want. A dummy load gives your system the ability to automatically apply a load to your turbine, which will keep it under control when your actual usage is not providing enough of a load for the current wind conditions. Your charge controller (see Chapter 11) will divert power into a dummy load when nothing else is using power, just to act as a brake on the turbine. The dummy load is not some mysterious black box that gobbles up all the excess power being generated, but it is something the charge controller can send power to. You actually *want* to waste all that excess power to create a load on the turbine and prevent it from spinning out of control. The best way to waste electrical power is to use it to create heat. Anything that creates heat through electricity is very inefficient, so the dummy load used by most charge controllers is a heating element.

We also need to discuss the optimal position for mounting the turbine. The higher you can mount it the better, so you will need a mast, a foundation for the mast, and supporting guy wires (also known as guide wires). You will have a bit of a construction project on your hands. Also, as in real estate, you need to consider location, location, location. If the tower comes down, you don't want to risk injury or damage. You should also keep in mind that wind turbines do make noise. You may come across discussions on rooftop mounting; we do not favor this idea, as you will suffer from too little wind and too much noise. When deciding on a position, consider rural farms that used windmills to pump water. Think of the height of those mills and their distance from any other structures that would interfere with the wind or be affected by the noise.

As with micro-hydro, we suggest that if you feel you have enough wind for a turbine to be a useful power source, it is time to find yourself a commercial vendor of wind turbine systems to evaluate your site and help you design a system. Be sure to consider the topic of repair and the wearing of parts. While a micro-hydro system is easily accessible to replace parts and make repairs, the same will not be true with a wind turbine located high up on a mast. See Appendix E for our suggestions on wind turbine resources.

# PHOTOVOLTAIC SOLAR POWER

There are many ways to use the energy from the sun. Heating objects, like cooking in a solar oven, heating water in a passive solar hot water heater, converting solar energy to electricity, and even hanging out your washed clothes to dry, are all examples.

When discussing the conversion of sunlight to electricity, we are discussing photovoltaic solar. "Photo" refers to the photons, or light particles, coming from the sun, and "voltaic" refers to voltage. When we mention solar, or solar electric, we are really discussing photovoltaic (PV) solar.

This alternative energy solution has a distinct advantage over the other two methods we have discussed because it has no moving parts. Solar systems (while we try and stay firmly rooted in reality, we do indeed have our own "solar system") do wear out, but not due to moving parts that have to be replaced. Instead, they wear out from age. But this happens at a good, ripe old age; most panels have 20- to 25-year warranties. Panels are strong, though you should consider some form of overhang that does not block sunlight but does protect from weather dangers such as hail.

Over the past decade, there have been many improvements in solar panels and as a result, there are many different types of panels. Another development in panels is a decrease in cost; you can now obtain panels for under $1.00 per watt. We will take a closer look at the different types of panels in Chapter 8 and consider their differences.

As of the writing of this book, we have been using 100 percent solar electric at our cabin for four years. We power all our lighting, computers, and printers; our computer network and Internet connection; our water pumps and fans; our TV, radio, and media players—all of this from electricity converted from sunlight. If you have deduced that we are firm promoters of solar electric as the choice for alternative power, you are right! Later in the book, we will lay out the steps for setting up a system similar to ours. You will learn how to determine the placement of your solar panels, the important considerations in constructing your racking (the adjustable frames for your panels), and how to take care of all your wiring needs. In most cases, preppers do not want to build a grid-tied system to sell power back to the utility company because their main goal is to be off-grid, but we will include some discussion of grid-tied systems in Chapter 15.

Unlike with wind and hydro, PV solar electric is an alternative energy option that you can set up independently. You do not need to contact a vendor or contractor for a site survey or system design and construction. More importantly, you do not need financing. Financing is useful for systems that are tied to the grid, where you are producing power to sell back to the electric company and lower your monthly electric bill.

The system we have built that meets all our needs mentioned above cost us $2,500. A big part of keeping the cost low was changing the way we use electricity to a low-voltage lifestyle, which we detail in Chapter 5: Watching Your Waste-Line.

# CHAPTER 2

# DON'T BE FUELISH

## Looking at Fuel-Based Alternatives

In our look at alternative power solutions, we have introduced three solutions that are self-renewing and do not require any continual supply of a fuel source. As long as the wind is blowing, the water is flowing, or the sun is shining, you can produce power. Now we'll take a look at alternatives that *do* require a fuel source, and the pros and cons of each.

The "two is one and one is none" rule tells us that fuel-based generators are good backups for wind, hydro, or solar. Our solar electric system with its battery bank can go for about six days of no sun. On day two, we become more cautious in our usage. On day three, we become frugal. On day four, we tighten our belts. Prayers are issued on day five. On day six, we pull out our backup and fire up the gasoline generator, plug in the battery charger, and connect it to the battery bank. We have only had to do this twice in the past year, but when we needed to, we were glad we had a backup plan.

Earlier we discussed how an electrical generator, either a direct current–producing dynamo or an alternating current–producing alternator, is basically a magnet rotating inside of a coil of wire. Though this electromagnetic process does not vary much, many different power sources can be used to rotate the magnet. We've looked at how to spin the magnet with wind and water flow. Now we'll look at using an engine.

First we must explain the difference between a motor and an engine. A **motor** is a machine that creates motion from electricity, such as a fan or an electric saw. An **engine** creates motion from the combustion of a fuel, such as a car engine that uses gasoline. If you go to the hardware store and buy an electric generator, odds are you bought a device that has an engine connected to an alternator. The alternator, in turn, probably produces 120 volts of alternating current to a built-in outlet that can power tools and other devices in your home.

# MECHANICAL OR INVERTER GENERATORS

You may not realize it from looking at a light bulb, but if it is running off standard 120-volt AC like most light bulbs do, it is turning on and off 120 times a second. With an alternating current, one wire is positive and the other negative, and at 120 volts, they reverse or alternate every 1/120 of a second. They do not do this abruptly, but rather, they start at zero volts (an "off" position) and charge to 120 volts, then decrease back to zero, and then repeat the cycle rapidly. Here is a picture of what it would look like if you could see electricity.

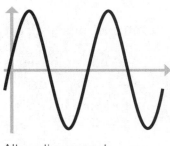

Alternating current

At 120 volts AC, this alternating current reverses its direction, or polarity, 60 times a second. This frequency of change in polarity is measured by a unit called hertz, and 120 volts AC alternates at 60 hertz. Heinrich Rudolf Hertz was a scientist in the 1800s who proved the existence of electromagnetic waves, so the basic unit of oscillation was named after him. But if our household AC power alternates 60 times a second, then why do our light bulbs turn on and off 120 times a second? Look again at the picture on page 21, and notice that in one complete cycle, the voltage crosses zero (the center line) twice, once on the way up and again on the way down.

Okay, enough of the science lesson. Let's move on to why this matters. All of the AC devices we use are designed for exactly 60 hertz. Give them any other frequency, and they will not run properly and may be damaged. Even our older clocks keep time based on that frequency.

When using your own generator to produce power, the speed at which the generator is spinning controls the AC frequency. If you've ever listened to a running generator, you know that it surges to different speeds as the load changes. The mechanical speed control of the engine does its best to maintain 3,600 revolutions per minute (60 cycles per second) but does not do this as well as the utility company. In most cases, the frequency is close enough (we call this "Good Enough Engineering"), but in recent years a better solution has become available in inverter-based generators.

Mechanical generator

Inverter generator

Rather than directly spinning an alternator, the newer design's engine does not use AC at all. Instead, it spins a direct current generator, which in turn powers an electronic device called an inverter. An inverter is a device that takes a direct current power source and converts it electronically into a highly regulated and controlled alternating current power source. Usually the low-voltage direct current is also increased; for example, you may own an inverter that plugs into your car's 12-volt direct current cigarette lighter to provide you with 120 volts of AC power to run your computer or TV.

There are two advantages and one disadvantage to an inverter-based generator. The pros are that you'll wind up with a lighter-weight, quieter generator with a well-regulated frequency and voltage. The con is that the design has more points of failure, meaning there are more ways it can break due to the added electronics. Though no one knows for sure, these may also be more vulnerable to **electromagnetic pulses (EMP)** or **coronal mass ejections (CME)**.

# FUEL

Gasoline, diesel, and liquefied petroleum (LP) are the three options for fuels that can be used to power a generator's engine. As we mentioned at the start of this chapter, none of these fuels are self-renewing. Once you run out of fuel, you have run out of power. So how much fuel you can store, how long it will remain usable (the fuel's shelf life), and how safe it is to store are all factors that will determine the usefulness of your generator when you can no longer procure fuel.

## GASOLINE

Gasoline is the most available choice of fuel, and most stores exclusively carry gasoline generators. Not only are these generators the easiest to find and the lowest in price, but gasoline itself is the easiest to purchase. There are two main concerns in storing gasoline: safety and shelf life.

While many off-gridders may be out in secluded rural locations (see Appendix A), others need to abide by local ordinances that control the amount of gasoline that can be stored on a property. Some locales limit your storage to as little as 2.5 gallons! And even if you are not controlled by local ordinances, realize that they exist for very valid safety reasons. Imagine what would happen to you, your loved ones, and your property if 500 gallons of gasoline were to ignite accidentally. The bulk storage tanks at gas stations are underground for a reason. Where will you be keeping your stored gasoline? Store it as far away from your residence as you can.

Also consider that gasoline has a short shelf life. You will need to store gasoline in full, airtight containers to avoid water contamination caused by condensation. Most gasoline today has a small percentage of ethanol added, and ethanol is hydrophilic, meaning it sucks moisture out of the air. For storage purposes, we suggest that you only purchase gasoline that does not contain any ethanol.

Over time, gasoline builds up varnish that ruins it and can damage engines. On average, you can only expect a three-month shelf life for gasoline. Products like STA-BIL and PRI can extend storage to two years. It is important to rotate your stored gasoline, use some on a regular basis, and replace it with new gasoline and more stabilizer.

You also must consider 2-stroke and 4-stroke gasoline engines. A **2-stroke engine** does not have an oil pan, oil pump, or oil lines that lubricate the engine like you would find in a **4-stroke engine**. Instead, oil is mixed with the gasoline, and as it burns in the engine, it lubricates the moving parts. Due to this simpler design, a 2-stroke engine is lighter. However, it also requires you to store enough of the 2-stroke oil to mix in. A 4-stroke generator is easier to own and operate.

# DIESEL

Diesel-powered generators are a bit harder to find, but worth the effort. Diesel generators have fewer parts. For example, they have no electrical ignition system that can fail, and so are better protected against EMP or CME. They typically have a longer life span due to their strong design. Most importantly, diesel fuel is much easier to store. We mentioned that bulk gasoline is always stored underground, but you may have noticed that farms and construction sites often have large aboveground tanks for diesel. This is because diesel is much less flammable than gasoline. However, you still need to be safe and keep it away from your residence.

Diesel degrades not only from water contamination but also from microbial contamination. Even algae growth is a concern. STA-BIL, PRI, and Dee-Zol Life make diesel-stabilizing products that should be used to help with storage. Some folks have reported that if they annually treat their stored diesel with the stabilizing products, they can use their diesel for 10 years or more. I read one report about 50-year-old diesel that was still usable. Fuel stations change their diesel blend for winter and summer; if possible, only buy summer blends to store. Winter blends produce less power in the engine, and due to the anti-gelling additives in the winter diesel, these blends can cost more. With all the things that can grow and build up in stored diesel, we recommend stopping at a farm supply store and getting a good diesel fuel filter, like a 20-micron filter, to pump the fuel from your storage container into your generator fuel tank. It is also *very* critical to be faithful and religious about oil changes. The best way to destroy a diesel engine is to miss an oil change.

One important consideration for both gasoline and diesel: When you are done with your generator, use the fuel cutoff valve to stop it, *not* the kill switch. If you use the kill switch, fuel will remain unused in the fuel system, which will damage the engine over time. If you instead use the fuel cutoff valve, the engine will quickly use up the fuel in the system and stall out. Having used up the residual fuel in the system, it won't jam it up.

## LIQUEFIED PETROLEUM (LP)

The best fuel to store is liquid petroleum, or LP. It is very safe. The only consideration you must make is protecting your tanks so they don't rust. Give them some type of a shelter.

LP's shelf life is unlimited. The only real downside with this fuel is that you need to have it delivered, which means you must go on the grid to have an account with an LP provider. You could use smaller portable tanks and fill them yourself, but LP generators use lots of fuel, and those smaller tanks won't last very long before they require refilling. As long as you don't blab that it is for your off-grid generator, they will think you are just like all their other customers and use LP for your furnace and stove. Remember your operational security (OPSEC) in your dealings with the outside world when buying your fuels.

So, where do you find an LP generator? Many farm supply stores carry them or can get them, as can some big box home improvement stores. But there is a more creative and flexible solution—you can convert a gasoline generator.

Unlike a gasoline or diesel engine that mixes combustible fuel with air sucked into its cylinders to create an explosive atmosphere, an LP engine simply starts out with an explosive atmosphere by breathing LP gas. Some companies make LP conversion kits to allow the connection of an LP tank to a gasoline engine. Should you ever want to go back to using gasoline, it is a quick disconnect to restore the original operation. See Appendix E for some resources.

# WOOD GASIFIER

During WWII, gasoline was hard to come by, so people got creative. If you have ever watched a campfire, you may have occasionally noticed flames dancing around on top of the wood, often looking like spouts or jets of fire. This is wood gas burning, and with a

Wood gasifier–powered vehicle

wood gasifier, it's possible to take advantage of the power stored in wood gas.

Obviously, making wood gas right in the fire is of no use; the flammable gas immediately burns up and does you no good. A wood gas producer consists of two compartments: a burner and a gasifier. One sits inside the other, and both are filled with wood. The lower compartment is lit, heating the wood in the upper compartment. As that wood heats up and starts to char, wood gas is released from the wood. This gas has lots of ash, dirt, and soot, which must be scrubbed out before entering the engine that will burn it. And once the wood producing the gas is used up, everything must be shut down to clean and refill. During that time, the engine has no fuel and stops. Also, wood gas is not produced at a high enough pressure to fill any kind of storage container; it has to be used as it is produced.

Check out YouTube or search online for "wood gasifier" for help with making your own. It's a great project, even if just for educational purposes. Another good resource is www.driveonwood.com.

# YOUR VEHICLE

For most of you, sitting in your driveway is a gasoline or diesel-powered vehicle that, with a simple addition, can power many of your electric needs. Your vehicle has a battery that can provide 12 volts of direct current. This battery is by no means the best choice for this purpose, but in a pinch, it will do the job.

An automotive battery is designed for very short-term, high-demand loads. It takes an immense amount of power to turn over your engine and get it to start, and an automotive battery can meet your engine's energy needs for short bursts. Once started, the engine of your vehicle also spins an alternator that recharges that battery. The battery can power lower demands, like your car lights or radio, for longer periods, but it is not designed for long, deep, repeated discharges like those that the batteries used for our battery banks are.

So, how do you use this vehicular generator? By connecting a power inverter to the battery. Some inverters come with heavy-duty clamps just for that purpose. Others come with screw down lugs that you can connect to cables. You can take an inexpensive battery jumper cable and cut it in half to make a great power cable for your inverter. Remember that in direct current circuits, the red cable goes to positive and the black cable goes to negative.

Inverter with vehicle cables

Inverter cables connected to car battery

Voltmeter showing battery voltage reading

Once the inverter is connected to the battery, plug a heavy-duty outdoor-rated extension cord into the inverter and run it to your house. There are many different sized inverters available for this solution, from 50 watts to 3,000 watts and higher. The larger inverters will need heavier cables to connect to the battery, and the more devices that are plugged into the extension cord, the quicker your vehicle battery will drain.

You don't need to keep your vehicle engine running the whole time, but you do need to keep an eye on the battery voltage so you know when to run the engine to recharge the battery. It's important to make sure that you have enough power left in the battery to get the engine started. Most inverters have a **low voltage safety disconnect** but often use too low of a voltage as the trigger. Many automatic disconnects don't trip until 10 volts, but I suggest that you not go below 11.9 volts on the battery before starting the engine to recharge it. A simple voltmeter will cost you less than $5.00 and makes it easy to monitor the battery voltage. Just set the meter for 20 volts DC (or the first range above 13 volts) and then connect the red and black probes to the plus and minus battery terminals. In the picture on page 28, you can see my battery is at 12.38 volts.

# CHAPTER 3
# WATT'S UP?

## Understanding the Nature of Power

We have work to do, so let's get started.

Having work to do is the reason this book exists. If we didn't have work to accomplish, we would not be in pursuit of alternative forms of power, because power is what helps us accomplish work. It might be backbreaking power, animal power, or the power from water flow, from the wind, or from the sun. When grid power is available to us, a lot of work can be accomplished.

Everything we've discussed in this book so far is a source of power, and for the most part, every source of power can accomplish the same work. Take, for example, the task of cutting a log: We can use a handsaw and our own human power to cut it. But to make the task easier, sawmills were built that used the current of a river over a waterwheel to turn a saw blade. There are designs for treadmills where livestock walk to turn the blades. Last weekend, I cut some boards with my cordless circular saw that had been charged from our solar electric system.

Since there are many sources of power and that power can accomplish work regardless of how it was created, power itself is universal. Because of this, there is a standard measurement for the power we have, need, or use, no matter where that power comes from.

# THE WATT

Power is measured in **watts**. The watt was named after James Watt, who greatly improved upon the steam engine, which in turn ushered in the Industrial Revolution. One of Watt's biggest contributions was developing a way to use the power of steam for many different jobs by inventing a means to produce rotary motion. Steam power could then be used to turn a saw blade, a millstone, or a pump, among other things.

Depending on the source of the power, different mathematical formulas are used, many of which are brain numbing. Fortunately, in this book we are focusing only on electrical power, so we have only one easy formula to deal with.

# ELECTRICITY

When explaining electrical power, a comparison to waterpower is often used, as it is easier to grasp and form a mental image. Let's set our task to sawing a lot of lumber. Imagine that we're going to build a sawmill powered by a waterwheel. As you can see in the drawing to the right, water flows down a pipe and pours over our wheel. The water hits the paddles on the wheel and pushes it around and around.

A waterwheel at work

What if we have a big, hard log to cut? Will our water-powered wheel be able to keep the saw blade turning, or will it bind and stop? If the blade does not have enough power to cut the large log, and it gets stopped by the log, how can we produce more power to get it spinning and cutting again?

To get the saw blade spinning again, we need either more water or faster water. Water turns the wheel in two ways: by the weight of the water on the wheel's paddles, as well as by the force with which the moving water hits the paddles. If we can deliver more water or faster water, then we can make the wheel turn faster and thus produce more power. To deliver more water we will need a bigger pipe, and to deliver faster water we will need more pressure.

Let's not forget our true topic here is electrical power. If we increase the amount of water hitting the wheel, we increase the wheel's power, which in turn increases the wattage. The same is true if we instead increase the speed of the water; we will increase the amount of power and the wattage. And there is nothing preventing us from increasing both the amount and speed, combining the power each method produces. The watts in this situation are the measurement of the flow times the pressure.

Now, let's put together our mathematical equation.

If we increase the amount of moving water, we have increased the current of the river. The word **current** is the same name we use to describe the flow of electricity. It is measured in **amps**. Grasp this idea and you'll understand half of our formula.

If we increase the speed of moving water, we increase its pressure. The pressure of electrical power is known as **voltage**, named for Alessandro Volta, the inventor of the battery. It is measured in **volts**.

So let's summarize *watt* we have learned (sorry, couldn't resist). The amount of power we need, have, or use is measured in watts. That power is made up of two parts: current and voltage. Current, measured in amps, is the amount of electricity flowing, and voltage, measured in volts, is the pressure behind that flow.

Now we can make our equation. In our formula, V represents the voltage, I represents the current, and W represents the total power, or wattage, produced.

$$V \times I = W$$

Here is an important aspect of this formula that many people struggle with, so we want to nail it down now. The mathematical operator in our formula is multiplication. Multiplication has a property called commutative property, which is a fancy way of saying the order of the numbers being multiplied doesn't matter.

$$2 \times 3 = 3 \times 2$$

This means that if 2 volts times 3 amps equals 6 watts is true, then 3 volts times 2 amps equals 6 watts is also true. If you increase the pressure (voltage) of electricity, you can lower the current (amps) and still get the same wattage. Or, you can decrease the volts and increase the amps for the same resulting amount of power.

## USING ELECTRICITY

Once you have obtained your electrical power, you need to be able to put it somewhere or otherwise make use of it. Perhaps you are producing power when the sun shines, or the wind blows, or the water flows and want to store it in a battery bank until it is needed. Or you may have an immediate need to use it: The refrigerator might need to cool things down, or it might be getting dark and some lighting is called for. You need to move your electricity from its source, or from where it's being stored, to where it can do some work.

As we just discovered, the watts stay the same whether you raise the voltage and lower the current or vice versa. This leaves us with the question: Which is better? Is there an advantage to either method of how to handle power? The answer is *yes*.

Electricity is moved over wires, and wires are the pipes in our waterwheel analogy. If you want to increase the amount of electricity being moved (in other words, increase the current), then you need bigger wires. Consider another example using water. If you are fighting a house fire, you want the large-diameter hoses that fire trucks use; you don't want to be using your small garden hose. The small garden hose is too constrained, causing too much resistance against the large amount of water flowing through it. Likewise, the thin wire has too much resistance against large flows of electricity. Compare this to the basic way that a fuse works: If you try moving more electricity through the fuse than it is designed for, the fuse will blow. And if you try moving too much electricity through a thin wire, it will get hot, meaning it will eventually fail.

If a wire is getting hot because too much current is flowing through it, then some of your precious power is being lost to heat. You are wasting power. This problem does not apply if you increase the voltage, which allows you to lower the current. That is why the power grid uses high-voltage lines to transport electricity across the country. The electric company raises the grid's voltage to tens of thousands of volts, which allows them to use thinner wires to transport grid energy. Once the power reaches its destination, the electric companies then lower that high voltage down to the 120 volts that most households use.

This ability to transport electricity at a high voltage is how Nikola Tesla and George Westinghouse beat Thomas Edison in the 1880s War of the Currents (see page 122). Edison refused to adopt the use of alternating current, preferring to stick to his direct current method, which he better understood. But when using direct current at that time, there was no means to raise or lower voltage. With the invention of the transformer, you could raise and lower the voltage of alternating current systems. Edison's direct current system could only transport electricity about a mile before the voltage loss in the cables made it unusable.

So learn the lesson that Edison had to learn: Higher voltage allows us to use thinner wires, as long as we have a way to reduce that voltage to a level that we can eventually use. This allows us to get the power we need, which is measured in watts, to accomplish our work.

# CHAPTER 4
# HOW TO EAT AN ELEPHANT

## How Much Energy Is Enough?

The answer to the old question of how to eat an elephant is "one bite at a time": piece by piece, and not all at once. In the previous chapter, we discussed how the power to do work is measured in watts. In this chapter, we are going to assess the work we plan to accomplish with our alternative power.

Most people that are connected to the electric utility grid have between 60-amp and 200-amp service.

Typically, 60-amp service is only found in older homes that have only 120 volts coming in from the grid.

Look back a chapter and recall our equation:

$$\text{wattage} = \text{amps} \times \text{volts}$$

This results in the ability to run a total load of 7,200 watts.

$$60 \text{ amps} \times 120 \text{ volts} = 7,200 \text{ watts}.$$

Newer homes with 100-amp to 200-amp service usually have two 120-volt lines from the grid, known as 240-volt service. The higher voltage allows for appliances such as electric clothes dryers, electric furnaces, electric stoves, and the like.

In the wiring for these houses, the circuit breaker service panel has some 240-volt breakers and circuits and some 120-volt breakers and circuits. The 240-volt breakers and circuits run the big appliances, and the 120-volt breakers and circuits run things like lamps, TVs, and more.

The total supplied wattage for these houses would be 24,000 to 48,000 watts.

$$100 \text{ amps} \times 240 \text{ volts} = 24{,}000 \text{ watts}$$

or

$$200 \text{ amps} \times 240 \text{ volts} = 48{,}000 \text{ watts}$$

These numbers reflect what your on-grid *capacity* is rather than what you are actually *using* on a regular basis. Which is a good thing, because your alternative power solution won't come anywhere close to those numbers. For simplicity sake, let's take the example from Chapter 2 where we discussed gasoline generators. A typical small portable gasoline generator can provide only 900 total watts. A moderately sized "roll around" generator, which is built on a wheeled cart, will provide a total of 4,000 to 6,000 watts. Some smaller generators have handles and can be picked up and carried around, but their smaller size produces 1,000 watts or less, making them less useful.

So as you can see, we are confronted with an elephant-sized job and only a mouse-sized solution.

Okay, it's not quite that bad...recently, we had some guests stay with us that would have had a miserable visit if I could not alleviate some of the summer heat we were experiencing. So I went shopping for small, efficient air conditioners. My concern was how much power the air conditioner required. The boxes they came in did not list the required wattage, but they did list two amperage ratings.

At first glance, this tag is confusing. Which is it, 15 amps or 9.8 amps? What the manufacturer is telling you is that the wall outlet that you plug this unit into should have 15-amp service, meaning that the breaker for this outlet should be a 15-amp breaker. But out of the available

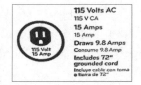

Air conditioner tag

15 amps, only 9.8 will actually be used by this air conditioner. We can find out using our familiar formula:

$$9.8 \text{ amps} \times 120 \text{ volts} = 1,176 \text{ watts}$$

This is what we needed to know. And in a perfect world, it would always be this easy. But alas, it isn't that easy, as the vast majority of manufacturers only tell you the needed capacity of the outlet, and not how much power will actually be used. Not all manufacturers provide such complete information on the shipping box.

So, we usually have to dig a bit deeper and look at the actual appliance. In most cases, you will find the needed information on the appliance very near where the power cord is attached. Often, the tag will tell you the required watts and no math will be required. Otherwise, just multiply the required amps by 120 volts to get the wattage.

| MODEL/MODÈLE NO. LP1015WNR | | |
|---|---|---|
| COOLING CAPACITY | 10,000 Btu/h | |
| INPUT | 1,080 W | |
| VOLTAGE | 115V | Hz | 60 |
| AMPS | 9.6A | PHASE | 1ø |
| DESIGN PRESSURE | H.S:450PSIG | L.S:250PSIG |
| REFRIGERANT | R410A 16.2 oz | |
| PORTABLE AIR CONDITIONER | | |

Power cord tag

Relatedly, it's time to learn a new term: volt-amperes or **volt-amps**, also expressed as **VA**. For our purposes, we are going to consider VA as essentially interchangeable with watts. The reason we mention it here is because some manufacturers put the VA rating on their tags rather than the watts, and we didn't want you to miss it.

Kill A Watt meter

As a last resort, if a device just does not display its power needs, you can use a simple wattmeter. A good one is Kill A Watt meter from P3 Company, available on Amazon for under $20. Just plug this meter into an outlet and then plug the device you want to measure into the meter. Turn on your device and read the watts being used on the meter. Read the instructions that come with the meter for more information.

Make yourself a list of the items you just can't live without (but we will try to change your mind about those in the next chapter). Remember that the goal is not to replace 100 percent of the power you use before the grid fails, but rather to be able to power the things you need to survive the outage.

Next to each item, write down its wattage. Now, make a note as to which items will be used at the same time, and total their wattage.

For example, in our home we need the lights at night, and we also "need" to catch the nightly world news. But when watching the news, we can turn off the lights because we don't need them to watch TV. And likewise, when we use the lights we don't need every bulb in the house blazing. We turn them off when we leave a room.

Look at your results, and pick out the highest total wattage load of the devices you need to use simultaneously. That is the maximum amount of power you will need to produce from your alternative power system. As you make your list, realize that some devices simply pull too much energy for any reasonable alternative energy system to power. You will not be able to run a washing machine, or an electric dryer, or an electric stove, or electric heating. Electricity-generated heat of any kind is a massive power hog and should be avoided at all costs. But there are many things that can run on an alternative power system and many ways to get your elephant-sized needs to fit within your mouse-sized capacity.

When calculating your energy needs, take into account how long a device will be used. Most alternative power systems will store power in a battery bank to assist during peak loads. For much of the time that your system is producing power, you will not be using it. Rather than produce excess power just to discard it, store it away in a battery bank (see Chapter 9). Then that stored power can be withdrawn to supplement the active power system when needed.

Let's say that you are building some wood shelves and need to run your power saw. Your saw needs 1,400 watts to run, but your alternative power system only produces 500 watts. But it has been running the past few hours, producing 500 watts, and you have only been using a 5-watt lamp during that time. That means all the excess power, 495 watts, has been going into the battery bank. Now when you need your saw, some of the power can come from your alternative power system and the rest can come out of the battery bank. You only use the saw for 30 seconds for each cut, perhaps a total of five minutes for all the cuts. The 1,400 watts is only needed for a total of five minutes, well within the range of what was stored in the batteries.

When we first went off-grid we only had 15 watts of solar power, yet we were able to use our computers, a light, and a connection to the Internet. We slowly expanded that system to where we now run an electric fridge, multiple lights, water pumps, a ceiling fan, multiple desk fans, a ham radio, computers, printers, television, and a DVD and media player. Our solar electric system produces a maximum of 900 watts in peak sunlight, and typically only 700 watts at any given time during the day. We have about $2,500 invested in this system.

In the next chapter, we will explain how to adjust your lifestyle to make all this work, like we did.

# CHAPTER 5
# WATCHING YOUR WASTE-LINE

## Changing Your Lifestyle

Four years ago, we made a major lifestyle change—we moved off-grid. At the time, we were living in a small midwestern town, we both had computer-related jobs that let us work from home, and we felt the desire and the need to establish a much more self-reliant lifestyle. So we let the research and decision-making begin.

As we began looking for property and planning our new life, we started prioritizing our goals. Alternative off-grid power was at the top of the list. From a prepper's standpoint, finding yourself in a long-term grid-down situation could be quite daunting; even those who have prepared will have a lot of work cut out for them if they are still dependent on the grid. In our case, we jokingly tell our friends that they have to let us know when the power is out so we can turn our lights off too. We love living our off-grid life, and we have been able to adapt to this way of life.

One thing that made the process easier for us was that we jumped off the grid right at the start. While the property we purchased had a house on it, the house had been vandalized while sitting vacant and all the wiring had been stripped from inside the walls. So we started at ground zero, with no power to the house, no wiring, no modern conveniences that we would have to wean ourselves from. It would have been harder to move off the grid while remaining in a house that had been connected to the grid and making ourselves give up those existing conveniences.

That being said, there were many changes in some of our everyday chores, and many common appliances disappeared from our home. It is not reasonable to expect your alternative power solution to equal what you would receive from the electric utility grid; you will have to adapt your power usage to meet your power production capabilities. The first thing to consider when trying to cut back on day-to-day power usage is that any appliance used to turn electricity into heat is going to use a lot of power. A quick glance around your kitchen will probably produce quite a list of such appliances—coffee pot, toaster, waffle iron, and slow cooker, to name a few. And then there are the power hogs in the bathroom, such as blow dryers and curling irons. These are just a few of the appliances we left behind in our old life. But the list doesn't end there. We also left behind major appliances such as our washer and dryer, furnace, dishwasher, and stove.

Based on the list of modern conveniences we left behind, you are probably thinking our life is pretty destitute, but that is hardly the case. We've adapted with plenty of non-electric solutions, and we can't discuss these without mentioning Lehman's Hardware. This hardware store in Kidron, Ohio, opened in 1955 to serve the local Amish population, and today they sell the same type of products for the non-electric lifestyle through their catalog and online (www.lehmans.com). While we do not live the non-electric lifestyle, we made a road trip to Kidron, Ohio, to pick up many of their alternative appliances before jumping off the grid.

Now we're going to take you on a tour of a typical day so you can see what some of our non-electric alternatives look like.

# A DAY IN THE SELF-RELIANT LIFE

First thing in the morning, of course, is coffee. We prefer freshly ground beans but left our electric grinder behind in favor of a hand-crank grinder that screws onto a mason jar, which collects and stores the grounds. We then boil water on our propane gas stove and make coffee in our French press, which requires no power. During cold weather, we can heat our coffee water on top of the wood stove. The coffee grounds steep in the hot water in the carafe of the French press. The lid of the press has a plunger that is slowly pressed down when the brewing is complete to push the grounds to the bottom, which separates them from the coffee. This past winter, we did pick up another alternative for hot beverages that Arlene was very happy about. She always enjoyed the variety her Keurig-style brewer offered with all the K-cup flavors, and there is now an off-grid version. It is basically a single-serving French press that uses K-cups.

Would you like some toast with your coffee? Again, we have a few options because we like to experiment. Our first stovetop toaster

Coffee grinder

French press with K-cups

was the typical camping toaster: the four-sided pyramid-style that always burns the bread at the bottom and leaves it untoasted at the top. A quick search on Amazon showed us several other options that gave us much better results—the Primus Toaster and Camp-A-Toaster.

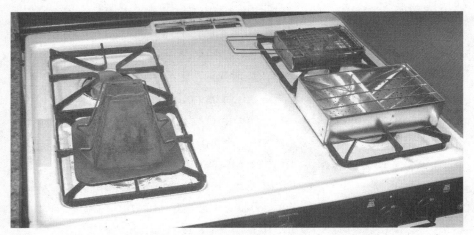

Options for making toast: traditional toaster (bottom left), the Primus toaster (top right), the Camp-A-Toaster (bottom right).

After breakfast, it is time to start a load of laundry. When you are off-grid, be prepared to discover that some chores will take a lot more time and effort without all the modern conveniences, and laundry is one of them.

1 We put our homemade laundry soap mixture into our laundry tub, which is a 17-gallon rope-handled durable plastic tub we picked up at Walmart. The laundry soap consists of equal parts of grated Fels-Naptha Soap, Borax, and washing soda. Do not use store-bought laundry detergent because it creates way too many suds during the churning phase.

2 Next, fill the laundry tub about one-third full with water and mix to dissolve the laundry soap. Add a load of clothes. Three or four shirts and two days of underclothes is a full load.

**3** Use an agitator like the Breathing Hand Washer, available from Lehman's Hardware, and "plunge" around in the tub about 200 times. You will probably need to build up your stamina to eventually do this all in one go.

**4** Remove the clothes from the washtub and run them through the hand wringer to drop into the catch basin. The hand wringer is also available from Lehman's Hardware.

**5** Place clothes in a 5-gallon bucket, rinse with clean water, and run through the wringer one more time. Take the clothes outside and place on the solar clothes dryer, more commonly known as a clothesline. We actually prefer to do laundry when it is raining so we can skip the rinse and second wring cycle and hang them outside in the rain to be rinsed. This saves effort on our part and keeps more water in our rainwater collection system.

Arlene demonstrating our laundry system and the Breathing Hand Washer.

The first time we did laundry at our property, we thought it was going to be a near-impossible chore to keep up with. But like several other time-intensive tasks, we picked up a routine and also dropped our preconceived notions of how much time it should take based on its on-grid counterpart.

Is it time for lunch? When we moved in, the house had no stove, but we did bring several cooking options: a countertop butane burner (think omelet station at a buffet), an outdoor grill, and a solar oven. After a few months, we found a used kitchen range on Craigslist that was set up for propane, which was just what we were looking for. It was also a very basic stove that required no electricity—no clock, lights, timers, or anything else. We wanted a propane stove rather than one that used natural gas so we could stay off the grid. Remember that the grid is more than just the power grid and that we do not want to be on any utility service or appear on any delivery schedules.

But wait, you may ask, what about propane delivery? We have all of our propane in 40-pound tanks because those are the largest we can carry ourselves and we take them into town to get them filled. Yes, at some point after the grid goes down, we will not be able to get them refilled, and then we will fall back to our other, more rustic cooking methods. But while it is available, we use the propane stove for cooking, canning, and as a backup dehydrator during overcast weather. On sunny days, we make plenty of power to run two electric dehydrators, but if we have had several cloudy days and the garden has produced an abundance that needs to be dehydrated, we can use the oven as a makeshift dehydrator. The oven pilot provides enough heat for most vegetables; you just need to remember to prop the door open slightly for the humidity to escape. To save on propane, we shut off the range-top pilots and use a lighter when needed.

Time to talk about the great non-electric multi-tasker in our living room. He actually works so hard all fall, winter, and spring that he gets the summer off—we're talking, of course, about our wood-burning stove. We never realized he would be such a multi-tasker, and yes, we admit that we refer to our wood burner as "he." That's because he is an integral part of the household from October to May. At one point early last spring, we took a picture of him accomplishing the following tasks all at the same time:

- keeping the house warm

- humidifying the air

- heating water for tea

- raising two loaves of sourdough bread

- keeping seed beds warm for germination

- drying clothes

- heating rocks to put in the worm farm

- keeping the cats happy

The multi-tasking wood-burning stove

Dinnertime, the last meal of the day, lets us look at one more cooking option: the sun. We use the Sun Oven solar oven, which you can read about in detail at www.sunoven.com. There are also many plans to be found on the Internet for building your own solar oven. Before we got ours, we were dubious as to how much of an oven it would really be, but we were soon impressed when it reached a temperature of 350°F. If you are going to be cooking anything for a long period, you will occasionally need to re-adjust the oven so it is directed toward the sun, and baking does take a bit of practice. We particularly like to use the sun oven as a slow-cooker. We can set it up in the morning, place a pot of ingredients in it, and then adjust it throughout the day to make sure it stays within the range of slow-cooker temperatures.

As we end our day of looking at some of the non-electric alternatives available to help lower power needs, it's time to relax with some hobbies. Arlene enjoys sewing, and she was able to pick up an antique treadle sewing machine. It came complete with an antique sewing machine, a cabinet, and a working foot-powered treadle system to operate manually the sewing machine. But, in line with our high-tech lifestyle, we found a company that manufactures current-day sewing machines that operate on the treadle system. We purchased this new machine and now have a modern sewing machine that runs on a foot-powered treadle system.

Sun oven

Modern sewing machine powered by a foot operated treadle

# CHAPTER 6
# POWER FROM THE PEOPLE

## Self-Made Electricity

You've seen us use the phrase "two is one and one is none," and it bears one more look. If your goal is to be self-reliant, then you should be striving to meet your needs without having to rely on outside resources. If you are a prepper, then your goal is to be prepared for as many potential problems as possible. If you are a survivalist, then your goal is to try to survive any eventuality, whatever might be thrown at you.

To do either requires redundancies and backup plans for when your primary plans fail. How will you meet your power needs when a hailstorm takes out your solar array? What will power your lighting and communications gear when your generator breathes its last and locks up?

Earlier we discussed how energy can only be converted, not made. Your own body's efforts to produce electricity result in rather poor energy conversion. Though it does convert matter into energy, the process of eating, digesting, fueling your muscles, and expending mechanical effort to generate electricity has many inefficiencies. The average adult of average fitness can sustain an output of 75 watts, while the other solutions we have looked at produce an output of 500 to 5,000 watts. That's why this is our last resort solution when all else has failed. We would much rather generate our power using the forces of nature, leaving us free to do other chores.

Let's take a look at a couple of solutions.

# K-TOR

The K-Tor Power Box is a 20-watt portable hand- or leg-powered dynamo that converts your physical efforts into 120 volts of direct current. It is designed to be used with your feet, which can pedal the device with less effort and for longer periods, but if you prefer you can use

The K-Tor Power Box

your hands. The Power Box has a standard AC-style outlet to plug your devices into. The 20 watts allow you to plug a USB adapter into the outlet and charge USB devices such as cell phones (even if the cell network is down, smartphones have many other uses), tablets, e-readers, rechargeable lights, and more. You can also plug in a 12-volt battery charger to recharge your main battery bank. We do wish that the Power Box had a 12-volt direct current output so that we could charge our battery bank without the inefficiency of having to use a 12-volt DC battery charger.

The K-Tor, being lightweight and portable, is designed to be screwed down when in use. Get yourself a comfy chair and start pedaling!

The K-Tor is available from k-tor.com for $195, and from our testing, it has every appearance of having a long life. It is a rugged device and folds down into a size that will fit in a bugout bag. When the grid is down, it will make the difference between sitting in the silent dark or having lights and communications. K-Tor also sells a smaller 10-watt hand-crank generator for $65. We were surprised by the number of references to the company in stories of how people bridged the grid-down situation during Hurricane Sandy in 2012. K-Tor has been working out of Vermont since 2008, so in 2012 during the hurricane, there were lots of people using these.

Alan catches up on some reading while using a K-Tor.

We need to discuss one issue with the Power Box, and that is its unique output. The wall outlets in a typical North American house provide 120 volts of alternating current. You may have noticed and wondered about a statement that we made a couple of paragraphs back, where we said that the Power Box outputs 120 volts of direct current. That was not a typo; this is a dynamo that produces direct current. On the face of things, there is not much in this world designed to run on 120 volts of DC. Most devices with motors, like fans, will not run on DC.

Wall wart power supply

Older clocks will not run on DC. And worst of all, anything powered off a transformer, like most of those little **wall wart** power packs that plug into the outlet, will not work.

What will work are any **switched power supplies**. A switched power supply is the newer version of those older wall wart power supplies. They use solid-state electronics instead of a

Brick style power adapter

transformer. Unfortunately, it is hard to identify this type of adapter just by looking at one. Common switched power supplies are USB chargers used by cell phones and many tablets; most laptop chargers are also switched power supplies. We have run our laptops from the Power Box with no problems. Adapters where the power supply is in line with the cord (known as brick power adapters), rather than directly plugging into the wall like you would with a wall wart power supply, are often switched power supplies. The bottom line is that you will need to try your device to see if it will work with the Power Box.

K-Tor sells a 12-volt battery charger for $18.00, which can be used to charge your battery bank. However, this charger only outputs 0.8 amps at 12 volts, and using our now familiar formula, $12v \times 0.8a = 9.6$ watts, we see that this is only about half of what the Power Box can produce. This will double the amount of pedaling necessary to

K-Tor charger

charge the battery than if the full 20 watts available could be used.

# PEDAL POWER GENERATOR

The next human-powered product we looked at is the Pedal Power Generator (pedalpowergenerator.com). This product is a stationary stand for a standard bicycle that has a 12-volt direct current dynamo mounted to the stand. A universal belt connects the bicycle's rear wheel, with the tire removed, to the dynamo. We picked up a bicycle at a thrift shop for $15 and we were off and running (er, pedaling) and charging our batteries.

The Pedal Power Generator can convert your pedaling into electricity, up to 300 watts per hour (200 continuous watts per hour, on average), for recharging your battery bank.

Left: Pedal Power bicycle adapter. Right: Alan using the Pedal Power Generator.

As we've mentioned earlier, a direct current dynamo such as the one in this product is an electric motor. If you spin the rotor in the dynamo, it outputs direct electric current. But if you connect direct current into the dynamo, it acts as a motor and spins. So what happens when you connect this product directly to your battery bank? If you guessed that it takes off spinning, turning your bicycle's wheel, and in turn, the pedals, you would be exactly right!

It is amusing to see a person try to leap on the bicycle, get their feet on the pedals, and get going fast enough so that the power starts flowing out of the dynamo and into the batteries, rather than the other way around. There is a solution: It is called a **blocking diode**. A diode is an electronics part that allows electricity to flow in only one direction.

Diodes are rated at various wattages, and you will need one with a high enough rating so that it can handle the dynamo's output. Fortunately, Pedal Power knows this and sells one. From looking at their website, it appears they have a newer model generator (PPG-B300W) with a built-in diode, which is a good improvement. If you obtain the older model, be aware that you will need to purchase the blocking diode separately.

Blocking diode

We have owned the Pedal Power Generator for four years now and have not had any issues. The all-metal (excluding the belt) product is extremely strong. The manufacturer also sells a "roller"-based product in addition to the belt-driven model. You do not want the roller version because it uses friction between the bicycle wheel and the roller to transfer the rotary motion, and in this case, friction is a bad thing. It is inefficient, and you'll tire out quickly. If you pedal harder or faster, it will also start skipping.

# PART TWO
# YOU ARE MY SUNSHINE: WORKING WITH PHOTOVOLTAIC POWER

# CHAPTER 7

# A TRIP AROUND THE SOLAR SYSTEM

## How Sunlight Gets to Your Light Bulbs

For a solar-powered system to work, the energy needs to travel from the panels to your devices. Before digging deeply into the science of the solar system, let's take a quick tour.

Briefly, photons from the sun are absorbed by solar cells, which convert photons into electricity. Multiple solar cells are connected together into panels, and multiple panels can connect to form solar arrays. The collective electricity from all these cells goes into a charge controller, which is responsible for storing the combined electricity efficiently and safely in the battery bank. A distribution system safely feeds the power to all the devices in your home. An optional inverter can convert the stored low-voltage direct current into 120 volts of alternating current if you have devices that require it.

Now, let's take a closer look at some of the component parts of a photovoltaic solar system. In the following chapters, you'll get to know each of these parts even better.

# SOLAR PANEL

In a photovoltaic solar system, the **solar panel** is the only part of the system that actually does the energy conversion to electricity. A panel, also called a **module**, is made up of lots of individual solar cells (see page 62), typically 36 to 72 cells per module. Smaller panels produce 5 to 15 watts of electricity, medium panels produce about 100 watts, and larger panels produce 250 to 300 watts. These are the **rated wattages**, or maximum wattages, as indicated by the manufacturers' testing. We like to say that these power levels were measured in a laboratory during a supernova of the sun; you will typically never see these levels.

A solar array is a collection of two or more solar panels connected together to obtain higher wattage. On an array that we tested, which had a projected output of 1,800 total rated watts, the highest output we ever measured was 1,600 watts. This is close to the rated 1,800, but 1,600 was the *most* we ever saw. For this array, the average output is 1,100 watts.

Solar array

All light, including sunlight, is made up of particles called photons. Solar cells absorb photons that strike them, and convert these photons into electricity. Even though they are called *solar* cells, this conversion process will work with any source of light; shine a flashlight on a solar cell, and it will output electricity. However, only the sun supplies us with vast amounts of photons without our having to take action to keep them coming.

# CHARGE CONTROLLER

The electricity from a solar panel can be used directly to power some devices, but as the amount of light striking the panel changes, there will be large voltage swings. You would not want to put up with your devices starting and stopping due to cloud cover, or the power level changing with the time of day. Many devices will need more power than the panels can produce during even the brightest sunlight, so they will need to be supplemented with stored power. It is the job of the **charge controller** to regulate the power from the panels and to see that power is stored for later use.

During the sunlit hours, the charge controller gets every bit of power that it can out of the panels and efficiently stores power into the battery bank without allowing any damage to the batteries. Some charge controllers can also protect the batteries from being damaged due to being drained too deeply.

To charge a battery, you must give it a higher voltage than it already has. Think about blowing up a balloon. As the balloon starts to inflate, you must continue to blow into it with high air pressure so it keeps expanding. Any lower air pressure will result in the balloon deflating and filling your lungs with air. But if you put too much pressure into the balloon, it will burst. The same is true with the batteries; the charge controller will provide the correct voltage at the correct times to fill the batteries efficiently and safely. The voltage coming out of the solar panels will vary as the sunlight varies, sometimes too low to charge the batteries, and sometimes too high to avoid damaging the batteries. The charge controller is responsible for controlling the charging voltage. For more on charge controllers, see Chapter 11.

# BATTERY BANK

If you only need electricity during the sunlit hours on sunny days and your panels produce enough wattage for the biggest power need you have, then you don't need batteries. Hopefully, you can see that this is not very realistic. A battery bank acts as a buffer, storing electricity until it is needed. When you use the electricity your system has

Battery bank

produced, you are limited by the combination of what your panels are producing at that very moment and what is stored in your battery bank. If it is dark out, only the electricity you have stored up is available.

You can build your battery bank to store electricity at a few different voltages, the most common sizes being 12 volts, 24 volts, or 48 volts. In the system we are discussing, a 12-volt battery bank will be used. For more information, see Chapter 9.

# DISTRIBUTION

The last part of the system involves getting your power to where you need it. The first step is to run the power through some type of protection, either fuses or circuit breakers. A fuse or circuit breaker is a device that breaks the circuit. If a dangerous amount of current starts to flow, it prevents electrical fires. There needs to be a master fuse or breaker that is large enough to power the total wattage you might use and still protect from a fire. In our home, the total maximum we feel we may ever use is 20 amps.

We will cover this in detail in Chapter 14, but it bears mentioning now that we run 99 percent of our home on 12 volts of direct current, not your typical 120 volts of alternating current. With 20 amps, we're using about 250 watts. After it travels through the main fuse or breaker, we send the electricity to a distribution panel

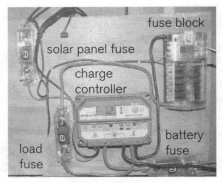

Example solar-power wiring configuration

that consists of individual fuses that power all the circuits in the house. We anticipate a maximum usage of 5 amps (65 watts) on each circuit, and a typical load of 2 amps (26 watts). To provide a bit of reserve, we generally use 10-amp fuses. If you were to add up all the 10-amp fuses for all the circuits, you would have a total way above the 20-amp main fuse, but remember that we don't have everything turned on at once, and most of the circuits are not pulling anywhere near their maximum. If we only used a single 20-amp main fuse instead of multiple 10-amp fuses on each circuit, we could easily have an overheating problem, especially if one the circuits designed for a maximum of 10 amps suddenly had 20 amps flowing through it. If 20 amps were to start flowing through one of the circuits, the 10 amp fuse or

Power inverter

circuit breaker on that wire would open the circuit and prevent the overheating.

Many of you will want to run your home on the 120 volts of alternating current that you are familiar with using on the grid. This will require one additional piece of equipment: a power inverter. This device will take your direct current battery voltage (12 volts in our example) and will convert it to 120 volts of alternating current. In the simplest design, your household devices plug directly into the inverter's outlets.

# COSTS

Let's end the tour by giving you a nutshell look at the costs involved. We've often said that if a conversation about a solar electric solution starts with the words "we can provide financing," then you are in trouble. The 12-volt system we personally use and are laying out in this book should not require any financing beyond a credit card, if that.

> Four-pack 265-watt solar panels . . .$1,200.00
> Charge controller . . . . . . . . . . . . .$668.00
> 4 batteries @ $85 each . . . . . . . . .$340.00
> **Total . . . . . . . . . . . . . . . $2,208.00**

Even with the miscellaneous wiring, fuses, and other parts, you are looking at a price under $2,500. If you choose to add an inverter for something that requires 120 volts alternating current (see our discussion of our refrigerator in Chapter 15), add in around $400.

# CHAPTER 8
# A PANEL DISCUSSION
## All about Solar Panels

In this chapter, we will be looking at the part of the solar electric system that actually gives us the electricity. **Solar electric panels**, also called **modules**, are made up of numerous (36 to 72) solar cells. Each cell produces half a volt of direct current. The amperage a cell can produce is determined by its physical size. Remember that power, measured in wattage, is volts times amps, so adding more cells increases the panel's voltage, and increasing a cell's size (which in turn increases the panel's size) increases the amperage.

A **solar cell** converts photons into electrical power through a chemical reaction. All light consists of particles called photons. When the material used to build a solar cell is hit with the photons that make up sunlight, the photons are absorbed and, through a chemical reaction, they release electricity.

Due to the low voltage (0.5 volts) that each cell can convert, cells are typically not found alone. And while a solar panel that is made up of multiple solar cells does produce a useful wattage at a useful voltage, they too are usually combined with multiple panels into an **array**, thereby greatly increasing the available total wattage to power your devices.

Since each cell can convert 0.5 volts, a 36-cell panel can output 18 volts and a 60-cell panel can output 30 volts. Most of the small panels that are commonly connected to a car's cigarette lighter socket to maintain a car battery are 36-cell panels; they are already a good size for directly charging a 12-volt automotive battery. If using a charge controller with these smaller panels, you would most likely use the simpler pulse-width modulation (**PWM**) controller (page 102).

The vast majority of larger panels, those rated at 100 or more watts, will have 60 to 72 cells per panel. As their output of 30 or more volts is much higher than the battery bank's 12-volt output, a multi-point power tracking (**MPPT**) charge controller is required. Often, multiple panels are connected in a series to simplify wiring, which raises the voltage even higher. We've seen the voltage on our system coming in at well over 100 volts, which is much too high to connect to anything but an MPPT charge controller (see page 101 for more information on MPPT charge controllers).

When you are looking at solar panels, you will find a number of different wattage sizes available. We suggest that you not use any size below 100 watts, with 250 or larger being preferred. You may also notice that many differently rated panels look alike, probably because they are built exactly like the higher rated panels. After construction, panels are tested for performance and wattage output and packaged based on those tests. It comes down to quality, not design, when ratings are determined.

# STANDALONE SOLAR CHARGERS

When we first went off-grid, we had an immediate need for electrical power and knew that it would be some time before we had our first solar electric system built. The quickest way to fill this need was with a few standalone chargers. Most of our immediate needs, including flashlights, radio, and USB phone chargers, could run on

AA batteries. We needed a few chargers for these batteries with built-in solar modules that we could set outside on sunny days.

This type of charger is available from a number of manufacturers. We purchased a few different ones for under $20 from Amazon. On page 62, we mention that more power requires larger solar cells. The physically larger the standalone charger, the faster it will charge the batteries. Pictured here are compact chargers, but in this case, smaller isn't really better.

Compact chargers

We still make daily use of handheld devices that use rechargeable AA batteries, so we have done a lot of research and comparison tests on the batteries themselves. We want

Eneloop battery

to share our results with you, which have shown us that Eneloop batteries from Panasonic are an excellent choice.

You may have heard that rechargeable batteries have a problem with memory that causes a battery's capacity to drop over time if they are recharged before being completely emptied; this was a problem with nickel-cadmium (NiCad) rechargeable batteries. That problem was solved with the introduction of nickel-metal hydride (NiMH) batteries, which are now the most commonly used rechargeable battery for handheld devices.

However, there are still a couple of disadvantages to NiMH batteries; namely, they have a short shelf life and fewer total charge cycles. A typical NiMH battery will lose up to 4 percent of its charge each day that it sits on a shelf or in a drawer. Compare that to an Eneloop battery, which will only lose 15 percent of its charge in a year. A regular NiMH battery will lose in three days what an Eneloop takes a year to lose. Additionally, typical NiMH batteries have a life span of about 400 discharge/recharge cycles, while Eneloop batteries are rated for over 2,000 cycles.

# TYPES OF PANELS

Solar cells have gone through many years of development, resulting in today's panels producing enough power at a price point where we can consider solar electric power to be a reasonable and cost-effective alternative power solution. Even a decade ago, solar panels were too weak and too expensive to be a solution for homestead, off-grid, self-reliance, or prepper use.

As panel production improved to where we could actually use it for our power needs, the technology that goes along with the panels also improved, and prices dropped. New charge controllers were developed that made better use of the panels and battery development has been progressing as well, though perhaps at a slower pace. While flooded lead acid batteries have not changed much, we have seen the development of AGM and Li-ion batteries (see page 87).

When looking at the development of solar electric power and when shopping for your parts, a useful measurement is dollars per watt. Look back a few years to see the change. We recently bought two 260-watt panels for $226 each, or $0.87 per watt. Back in 1977, solar panels cost $78 per watt, and as recently as 2004, they cost $5 per watt!

Throughout the history of panels, three main types have come into use: monocrystalline, polycrystalline, and thin film. The vast majority (over 90 percent) of solar panels are built of silicon. The main difference in types of solar cell construction is the purity of the silicon.

## MONOCRYSTALLINE

**Monocrystalline** panels have a high **purity**. By purity, we are not talking contamination, like pure mountain water; rather, we are referring to how well the crystals line up with each other. To make these cells, the silicon is formed into a cylindrical ingot, and then wafers are sliced off.

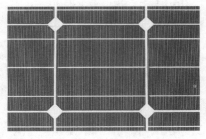
Monocrystalline cell

Due to this process, the solar cells have rounded edges, which is another way to identify them.

The purer, more perfectly aligned the silicon crystals, the better the cell will be at converting solar energy into electricity. Monocrystalline cells are 15 to 20 percent efficient. When you look at the cells in a monocrystalline panel, the cells look even in their coloring. They also have a longer life span and typically come with 25-year warranties. Being more efficient, they work better in low light conditions, such as on cloudy days. Because of these benefits, monocrystalline panels are also more expensive than polycrystalline or thin film panels.

## POLYCRYSTALLINE

This type of cell is an older, simpler cell to produce. The silicon is melted and poured into square forms, resulting in crystals that are not well aligned with each other and reducing the efficiency of the cell. **Polycrystalline** cells are 13 to 16 percent efficient. Though this process does come with a lower production cost and produces

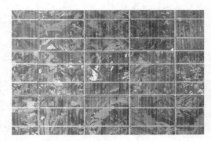
Polycrystalline cell

less waste, a polycrystalline panel will produce less power than a monocrystalline panel of the same dimensions. These cells are also more sensitive to heat, which will lower both their power output and their life span. If space for your panels is not a major concern, you might want to consider polycrystalline panels when you are shopping around.

# THIN FILM

**Thin film cells** do not use silicon, as mono- and poly-cells do. Rather, one or more thin layers of photovoltaic material are deposited onto a surface that forms the panel, almost as if they were spray-painted on. Thin film panels can be made from different kinds of photovoltaic material. No matter what kind is used, the result is a cell that has an overall lower efficiency than mono- and poly-cells, but is very inexpensive to manufacture.

Thin film cells are uniform in appearance and less sensitive to heat and to shading. Their life span (only 5 to 10 years) is much shorter than those of mono or even poly. If you have the space for them, thin film panels can save you money; being less efficient, it takes two to three times the amount of thin film panels to produce the same power as mono or poly.

Compare the two pictures on page 68. The one on the left is our original 270-watt array built from Harbor Freight 45-watt thin film kits. We started with these, as they were the lowest price per watt we could find. Each kit of three 15-watt panels cost around $130. The right-hand picture shows two of the Grape Solar 265-watt monocrystalline panels that we bought from Home Depot for $370. One of those Grape Solar panels produces almost the same amount of power as our 30-foot array of the Harbor Freight thin film panels. That's a big space difference, not to mention the cost difference ($2.88 vs. $1.39 per watt). This is part of what we meant when we said that one of our goals was to help you skip over the mistakes we made in our journey. We wish that these Grape Solar monocrystalline panels had been an option when we started out!

Harbor Freight thin-film panels    Grape Solar monocrystalline panels

# MOUNTING

Solar panels need to be safely mounted, which is called racking. Many people mount their panels on their roof, which we are going to try to convince you *not* to do. Roof-mounted panels may look nicer, but they have a number of disadvantages. The panels need to be re-positioned at least twice a year as the sun changes its position in the sky. Repositioning the panels requires you to mount them to a movable frame, and then the frame itself has to be mounted. Moreover, mounting the frame to the roof will punch holes in the roofing. It will require care to seal so that you don't create leaks in your roof. When it is time to re-position your frames, you will need to get up on the roof to do so. This may not be too hard of a task, but it is harder and more dangerous than if the panels were ground-mounted. Also, if you have snow in your area, you will find it much easier to remove from your panels if they are not up on the roof. We really cannot think of an advantage to roof mounting.

So, if not on the roof, then what will you mount them to? The answer is a solar rack. A rack can be made of metal or lumber and is typically supported on a post or posts that are cemented into the ground. As in all forms of ground construction, you need to make sure you get below your local frost line so that ground expansion from freezing does not move your rack and panels, especially in the parts of the country that experience freezing temperatures. In our situation, we have a deck on the south side of our house, so we built our racking out of lumber that is secured to the support posts of the deck.

To reduce power loss, you will want to be able to place your panels as close as possible to where your charge controller and batteries will be located. Try not to exceed a distance of 50 feet, so go look around your property and see if you can find an area with good southern exposure that gets sunlight without major obstructions during the daylight hours. Next, see if that area is within a reasonable distance from where you can place your equipment. If so, you will be done with your site survey and can move forward.

We were able to place our panels in an area that worked well, but after the first year of watching how the position of the sun moved over the seasons, we did end up removing a dozen trees, which are now firewood. The removal of the trees opened up the canopy, which allows for more hours of unobstructed sunlight. As you look around for a good location, keep in mind that you might need to remove some trees.

Solar panels are enclosed in a metal frame, which is what will connect to your racking. Panels have a tremendous amount of surface area that can be caught by the wind, so make sure you tightly secure your panels to the rack. A panel taken by the wind is not only an expensive disaster but a dangerous one that can cause death if it hits someone. Make sure your rack is structurally strong, well secured, and able to take the wind hitting the panels.

We mentioned that the panels have to be repositioned during the year as the sun moves. A buddy of mine built his racking in a way that allows him to crank the array to a different angle each month. Another buddy designed his racking so it can easily move into two positions, one for the vernal equinox and one for the autumnal equinox.

On the two equinoxes, the sun is at the midpoint of its arc across the sky. On approximately March 20, the vernal equinox, the sun passes through this midpoint on its way to its highest point, and around September 22, the autumnal equinox, it again passes the midpoint on its way to its lowest point in the sky. The sun reaches the highest and the lowest points on the solstices.

You can, of course, move your array every month, following the sun in its travels, and get the highest power level possible. We have chosen to go for the average and move our array twice a year. Compared to moving the panels monthly, we have not found much difference in only doing it twice a year. If you reposition every month, you will be pointing your array at the angle that points to the path of the sun. If you only re-position twice a year, you will point your array halfway between the midpoint of the sun's path and the highest or lowest point. These angles are different depending on where you are located. You can use this online tool to determine the sun's angle for each month, as well as the equinoxes: www.solarelectricityhandbook.com/solar-angle-calculator.html.

Let's use Washington DC as an example to compute the sun's path through the sky. The tool shows that you'd want your panels to have a winter angle of 28°, a summer angle of 74°, and an equinox angle of 54°. Using these numbers, at the vernal equinox we would position our array halfway between 74° and 54° for an angle of 64°, and at the autumnal equinox, we would position the array halfway between 54° and 28° for an angle of 41°.

If you wish to reposition your array each month, we have a quick and easy way to position the array correctly with just a nail. Take a common flathead nail, and glue the head of the nail flat to the frame of one of your panels, with the nail pointing at the sky. Wait for local noon, when the sun is its highest in the sky. The nail will cast a shadow on the panel. As you raise or lower the panel, the length of the shadow will decrease and increase. Position the panel so the nail shadow is at its shortest.

An example of using the nail test to correctly angle a panel

Here are some racking examples to give you an idea of ways you can mount your array.

Metal racking with monthly repositioning

Wooden self-supporting racking

Wooden deck-mounted
racking awaiting new panels

# GET WIRED: SETTING UP SOLAR ARRAYS

Let's first talk about the connectors that are used in solar electric. You are likely to come across two types of connectors: the 2 pin quick disconnect harness or the MC4 connectors (on next page). You are most likely to end up with the MC4 connector, unless you are using smaller kits. Small panels intended for direct automotive battery maintenance often use the 2 pin quick disconnects. The panels we are working with and discussing in this book will be using the MC4 connectors.

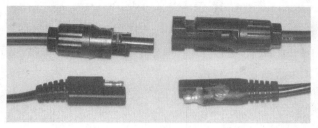

MC4 connector (top) and quick disconnect (bottom)

As we mentioned in Chapter 8, each of your panels has a rated wattage, which is a manufacturer's rating of how much power can be converted in bright direct sunshine. These ratings are very optimistic, as you will rarely reach the output predicted by the rated power of your panels. However, these ratings do allow you to compare panels and design your wiring.

Remember that power equals voltage times current. If the wattage of your total array is 1,000 watts and each panel produces 30 volts of electricity, you have two delivery options for that power. You can either deliver that 1,000 watts to your system as 30 volts times 33 amps by wiring the panels side by side in a parallel array (**parallel wiring**), or you can deliver it at 120 volts times 8.3 amps by running each panel's output into the next panel, adding up their individual voltages in series (**series wiring**). Here's an easy way to remember the difference: parallel wiring adds the currents together, and series wiring adds the voltages together. We'll describe this process in depth in Chapter 9.

It is important to remember that between the array and the charge controller, you need to include either a fuse or a direct current circuit breaker. When wiring a parallel array, each panel will need a fuse or breaker. For a series array, you can use a single fuse or breaker for the entire setup. The wiring coming out of each panel is usually rated for 30 amps, and each panel typically produces less than 10 amps. Once all the panels are connected to the combiner box (the busbars and a fuse/breaker for each panel are often contained in their own box), their combined currents flow over the wires to the charge controller and produce a higher current than the basic cables are designed for.

# THE SHADOW KNOWS: PARTIAL SHADING CONSIDERATION

We've just discussed wiring the panels that make up your solar array in series or parallel, and later we will discuss wiring that combines both series and parallel. A big part of deciding how to wire your panels is influenced by shading. As an example, let's consider a two-battery flashlight. If your batteries are dead, and you only have one fresh battery available, how well will that flashlight work if you only replace one dead cell? The dead battery will actually block power from flowing.

A flashlight with series batteries will not work if even a single battery is dead.

The same is true if you have a number of solar panels wired in a series and one or more of those panels are in shade. Your array will not start producing power until all of the panels have sunshine on them. If you instead used parallel wiring on the panels in your array, then the array will produce full voltage even

Panels in shade

in partial shading. However, it won't produce full current until all the panels have sunshine. With a parallel array, you won't get full power until all the panels are in sunshine, but some power is better than no power. As mentioned previously, the downside is you will need more wiring for a parallel array.

Before you lay out your array and before you decide on using series wiring, look for any partial shading problems. A wide array will bring up more potential shading problems than a narrow one. We sit in a valley, with one mountain on our east and another on our west. As the sun rises over the eastern mountain, the west side of our 30-foot Harbor Freight array gets light, but it takes a few hours for the entire length of that array to receive sunlight. Then as the sun sinks over the mountain on our west, the west side of our array that first received sunlight starts to lose it. This array, due to its partial shading problems, is wired in parallel. Our much narrower Grape Solar array is wired in series, as it has much less of a shading problem.

# CHAPTER 9
# SALT AND BATTERY

## Learning to Use a Battery Bank

As was mentioned in Chapter 7, a battery bank is not completely necessary to a solar electric system, but personally, we would not design a system without a battery bank. The battery bank acts as a buffer for those occasions that you need more power than your solar panels are producing at the time. For an example, let's look at the biggest energy hog we use in our home: the microwave oven. In the 12-volt off-grid system we are discussing in this book, the solar array can produce 1,000 watts under optimum conditions. Our microwave oven pulls a little over 1,400 watts, so it is easy to see that it would not run on just our solar electric panels.

Most of the day, our array is producing more than we are using. On average, we use about 50 watts at any given point during the day, though this amount increases at night, as the lights and television come on. So what should we do with the extra wattage (1,000 produced – 50 used = 950 watts excess)? The best idea is to bank those extra watts for when they are needed later on.

To be able to run our microwave oven, the power coming in from the solar array is added to stored power from the battery bank. With the excess power, we have enough to run the microwave oven. This would not be possible without a battery bank.

Hopefully, you can see that this extra power cannot be used forever: The battery bank starts to drain, and eventually will be exhausted.

A draining bank is most noticeable after the sun goes down, as 100 percent of your power will come from the battery bank. A lack of batteries after sundown will leave you sitting in the dark. Remember the one is none rule and keep some lanterns and candles around, just in case.

A battery bank is only as good as its worst member. This means that all the batteries in your bank live and die together. If a single battery fails, you should not simply buy a replacement and swap it out, as the new battery's performance will be greatly impacted by the rest of the aged bank. It will not get a full charge, and will not provide power to its best ability. Battery banks need to be built and put into service as a single piece. If one battery fails, it is time to replace the entire bank.

This rule is also true when expanding a bank. If you have a bank consisting of two batteries and a year later you decide to double it to four batteries, you shouldn't add two new batteries alongside the two old batteries. Rather, you should pull the older batteries out of service and replace them with four new batteries.

A few words of caution: You should *never* buy used batteries; if they were good enough to use, the owner would have kept them. And finally, never mix different sized batteries, and always match your batteries to be the same brand and model. All of your batteries need to be the same voltage and capacity, and roughly the same age. Design wisely, and do it right from the beginning.

# ABOUT THE BATTERY

In 1800, Alessandro Volta (for whom, you may recall, the volt is named) invented the voltaic pile, credited as the first battery (though some evidence has been found of batteries dating back to 200 BC). Industry through the ages has greatly improved on Volta's design, but the basic principle remains the same.

At the basic level, a battery consists of plates of lead submerged in acid (or in more technical terms, a lead electrode is submerged in an acid electrolyte) to create a chemical reaction, which causes it to produce electricity. Increasing the surface area of the plates by making them thinner results in the ability to draw more immediate power. Thicker, heavier plates allow for a longer lasting power draw.

Today, there are three types of batteries.

| TYPE OF BATTERY | DESCRIPTION |
| --- | --- |
| Starting Battery | Bears high load; has short duration |
| Deep Cycle Battery | Bears low load; has long duration |
| Accessory Battery | Bears moderate load; has moderation duration |

Starting batteries have to supply over 100 amps for five to 30 seconds to start your car engine. Deep cycle batteries run for hours on a smaller load; for example, these would be used for running your boat's trolling motor or for traffic signals. Accessory batteries run your cordless drills and saws, cell phones, tablets, or laptops. The functions of a battery are determined by its plate thickness, size, and chemical composition.

The deep cycle battery is best suited for our uses. We won't typically be drawing extremely large loads, and we want the power to be available for as long as possible. We want to be able to get from sundown to sunup without running out of power.

# BATTERY RATINGS

Batteries are rated either by **cranking amps** or by **amp-hours**. Sometimes a battery is rated using both measurements. We recommend staying away from any battery rated in cranking amps, as the manufacturer is implying that the battery can be used to start an automotive engine and is therefore not as well suited for deep cycle use.

Amp-hours tell you how long the battery will provide power. It is a misleading rating because it varies by how fast a battery is drained. A battery being drained by 1 amp might last 100 hours, but this does not mean that the same battery having 100 amps pulled from it will last one hour. It is not a linear relationship. The faster a battery is drained, the less overall capacity is available. The amp-hour ratings for most deep cycle batteries are based on a 20-hour period; they show how many amps can be pulled from the battery within 20 hours. This is called the C20 rating, which tells you how much current is available over a 20-hour drain. When you are comparing batteries for purchase, compare the C20 ratings between your choices.

# HOW BIG IS BIG ENOUGH?

Pretend your batteries are buckets. During the sunlit hours, you want to take all the electricity you produce and don't immediately use and store it in your battery buckets. Once the buckets are full, any excess power from the solar panels is just discarded and wasted. When the sun goes down, the power you put in the buckets has to last you until the sun again shines. If you are having a string of bad weather days (rain, snow, clouds), then the buckets may need to last a few days. We found that with our typical power usage, our battery banks will last us five days of overcast weather before we have to fire up the backup generator and battery charger to keep us going.

So you can see that the size of your battery bank is important. You may be thinking that you now have to do a bunch of math to figure out how much power you need and for how long. Fortunately, there are not a lot of choices. Your bucket size—battery size, really—will be dictated by the types of buckets available on the market.

Before we can continue, we need to revisit series and parallel connections. As a reminder, these are two ways to connect multiple batteries together when building a battery bank.

To make a series connection, take two 6-volt batteries and connect the positive terminal of one to the negative terminal of the other, like a freight train is connected in a line. Once you've done this, the voltages combine but the current remains the same. For example, imagine connecting two GC2 (GC stands for golf cart) batteries and

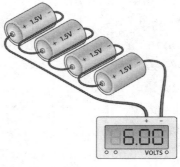

Series wiring

connect in a series, one feeding into the other. Each is rated at 6 volts and 208 amp-hours. When connected together, they will be producing 12 volts at 208 amp-hours: double the voltage but the same capacity. This is known as a **series string**.

For a parallel connection, take the same two batteries and lay them side by side. Connect the two positive terminals, then connect the two negative terminals together. You'll find that the voltage will remain the same but the current capacity will double. The result would be 6 volts at 416 amp-hours.

Batteries are made up of cells. A 6-volt battery has three cells and a 12-volt battery has six cells. You can count the watering caps on top of a battery to see how many cells it has. To build a 12-volt battery bank, you will need a minimum of two GC2 6-volt batteries in series. This will give you a total storage capacity of 208 amp-hours. If you were using 50 watts an hour, a bank of this size would give you about two days of power in cloudy skies.

So, you want more capacity, eh? Okay, get yourself two more of the GC2 batteries (but remember, you cannot mix old and new batteries), connect them together in series to make 12 volts, and then connect your two pairs together in parallel. Now you have 12 volts at 416 amp-hours. This formation is known as two parallel strings of two series batteries.

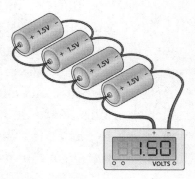

Parallel wiring

Not quite enough yet? Well, you are starting to reach your limit now. We recommend not going over two parallel strings because two will be well balanced when charging and discharging. Batteries have slight internal differences, which create internal resistance. As you add additional parallel strings, these differences cause problems with equal charging and discharging, which will shorten the life of the batteries. We won't come over and throw rotten vegetables at you if you decide to grow to three parallel strings, but that is the limit. A maximum of two strings is best, three is somewhat acceptable, but more than three is a very bad idea.

You can now see that your choices when using the GC2 batteries are 208 amp-hours, 416 amp-hours, or 624 amp-hours. Let's see how these choices match up to your ability to fill these batteries.

In the 12-volt system we are describing in this book, we have four 265-watt solar panels, so at their best performance, we can theoretically produce 1,060 watts in perfect conditions. At sunrise, only a little sunlight hits the panels, so little power is produced. As the sun climbs in the sky, more sunlight produces more power, continually increasing toward the maximum of 1,060 watts. If a little black rain cloud drifts by, the power will be reduced for a few minutes. Then as the sun starts sliding toward the west, power starts to drop off. Over the course of our daily period of sunshine, our 1,060-watt array of solar panels collects around 3,000 watt-hours of power. To put this into perspective, say the sun shines on our panels from ten in the morning to four in the afternoon. Therefore, we collected our 3,000 watt-hours of power in six hours, which averages out to 500 watts per hour. This means our array, which is rated as at 1,060 watts, actually performed at 500 watts in the real-world conditions.

We typically use 50 watts per hour, or 1,200 watt-hours in 24 hours. This gives us an excess of 1,800 watt-hours we can use to fill our batteries.

$$\begin{array}{r} 3{,}000 \text{ total watt-hours harvested} \\ -\ 1{,}200 \text{ watt-hours used} \\ \hline =\ 1{,}800 \text{ watt-hours remaining} \end{array}$$

To look at this from another perspective, let's convert watt-hours, which we use to measure energy production from our panels, to amp-hours, which we use to measure our battery bank. Remember that watts divided by volts is equal to amps. Taking our excess 1,800 watt-hours of produced power and dividing it by 12 volts, we wind up with 150 amp-hours. With this amount, we would not have enough excess power to fill our 208 amp-hour battery bank if it were empty.

The good news is that we never want our battery bank to be completely empty, and it takes fewer excess watt-hours to fill a partially empty battery than it would to fill a completely empty battery. Keep in mind that the inability to fill your battery bank regularly will decrease its life span. The bottom line is, don't oversize your battery bank to where your solar array can't regularly fill it, or you will damage your batteries. You want to be able to fill your battery bank fully at least twice a week.

You can always add to the size of your solar array to produce more power. Unlike adding to your battery bank, you can add solar panels to your array at any time. If the day ever comes that you have a large solar array and need to build a battery bank that stores more than 624 amp-hours, you can find 6-volt batteries with a capacity larger than 208 amp-hours each. These are simply bigger batteries. When you are at three strings, you may not be able to add any more batteries, but you can swap for bigger batteries. A friend of ours is using 415 amp-hour batteries compared to our 208 amp-hour batteries. These are, of course, more expensive.

So far, we have been discussing how many amp-hours we can store. But remember that our ability to do work and power our devices is measured in watts. Here's a reminder:

amps × volts = watts

Our 208 amp-hour battery bank stores 2,500 watt-hours.

208ah × 12v = 2,500wh

If we cannot increase the amp-hours of our battery bank because we have reached the allowed limit of two or three parallel strings, we *can* increase the voltage. A 24-volt bank will have four batteries in each series string instead of two.

4 × 6v = 24v

This will result in 5,000 watt-hours. We will look at a 24-volt bank more in Appendix D.

208ah × 24v = 5,000wh

# TYPES OF BATTERIES

## WET CELL BATTERIES

**Wet cell**, or "flooded lead acid," batteries are so named because the battery case is filled with a sulfuric acid, a liquid electrolyte. These are types of flooded lead acid (**FLA**) batteries. The charged lead plates, called **electrodes**, are inserted into the liquid electrolyte. Tip them over and they spill. Shake them and they splash. Mistreat them and they will either bite back and hurt you, or they will die.

While these types of batteries are charging, explosive hydrogen gas is produced (remember the Hindenburg?). Because of this danger, many folks keep their batteries outside their homes with plenty of ventilation. While ventilation is important, batteries do not like the cold, so just sticking them outside without protection is a great way to damage them. They prefer a temperature range between 32 and 80°F.

Ventilated battery box

Batteries need to be in a well-ventilated battery box so they are protected from rain, snow, and cold. Also, be sure to keep your batteries clean, ventilated, and away from flames or sparks. Our box is located in the utility room inside our house. It has a vent tube leading outside and a cooling fan from

Ventilation fan draws in air

a personal computer case that draws air into the box, which then pushes the hydrogen gas outside.

To ventilate your battery box, do not use any old fan you might have lying around. Most fans use older motors with internal parts that spark, which could detonate the hydrogen gas. PC cooling fans use induction motors with no sparking parts. Position your vent tube near the top of your battery box, because hydrogen gas is lighter than air and will collect at the top of the box.

Flooded batteries are the least expensive choice for your battery bank but require the most maintenance. We use Energizer GC2 6-volt, 208 amp-hour golf cart batteries we purchased from Sam's Club for $85 each. GC2 is a common size designation for golf cart usage.

Energizer golf cart battery

Golf cart batteries are the best-priced choice for small systems because they are designed for long duration, low current loads like those found in typical off-grid homes. Other reliable brands include Trojan, Rolls, and Fullriver. They will stand up to a lot of abuse, including undercharging. We are designing a 12-volt based system in this book, and since GC2 batteries are 6-volt batteries, we will be connecting two together in series to add up to 12 volts. Technically, each cell in a flooded wet cell battery produces 2.23 volts, so a fully charged 6-volt battery will measure 6.69 volts and a 12-volt battery will measure 13.38 volts. We mention this to avoid any confusion when you get your voltmeter out, but for simplicity's sake, we will continue to discuss batteries in terms of 6 volts and 12 volts.

Compare the size of 6-volt deep cycle batteries, and you will see they are almost as big as, and much heavier than, a 12-volt automotive battery. This is because the lead plates are so much thicker. We use

6-volt rather than 12-volt deep cycle batteries, because the thicker plates required to make a 12-volt deep cycle battery would just be too large and heavy. If you see a "deep cycle" 12-volt battery that is about the same size as a 6-volt deep cycle battery, you know that the plates are just not as thick in the 12-volt battery, thus making it inferior in storage capacity compared to batteries with thicker plates.

We'll be discussing the maintenance and monitoring of wet cell batteries in the next chapter, but first, let's look at some of the low-maintenance alternatives available on the market.

## I'M GELLING: GEL CELL BATTERIES

Gel cell batteries are very similar to typical flooded batteries, except the liquid electrolyte is replaced with an acidic gel. By using gel, the battery cannot splash or spill if tipped or shaken. These are maintenance-free batteries. You do not need to monitor the fluid levels; in fact, these batteries are sealed so you cannot access the gelled fluid. While it is true that all batteries off-gas hydrogen when charging, very little gas is produced while charging gel batteries and simple ventilation holes in the battery box will keep them safe.

Gel cells are easily damaged by overcharging. While flooded batteries just boil if you overfeed them, gel cells will be overpressurized and the safety valve will blow. Overfeeding may even cause the case to crack. When gel cells are damaged through overcharging, the battery will no longer be usable. Gel cells thus must be charged carefully.

Compared to flooded cells, gel cells have a shorter life span, a much higher price, and a lower capacity. You are paying for convenience, not for performance. Due to the high cost and the low performance, we do not recommend this type of battery for your off-grid battery bank.

# WELCOME MAT: AGM BATTERIES

**Absorbed glass mat (AGM) batteries** are a better choice than gel cell batteries if low maintenance and minimal off-gassing are important to you. These batteries contain glass mats that absorb and hold the acid. Like gel cell batteries, they are sealed to reduce spillage and do not need periodic filling. Priced between flooded batteries and gel cells at $450 for 415 amp-hours, AGM batteries are the first choice for many people.

As a plus, AGM batteries are often marketed for solar power use so they are available in large-capacity sizes. They are not as sensitive to charging rates as gel cell batteries and are charged in a similar method to flooded batteries. They are also less affected by cold temperatures than gel cells, so they can easily be stored outside in a battery box. They produce minimal off-gassing, so simple holes in a battery box will be sufficient for ventilation.

# THE LION SLEEPS: LITHIUM ION BATTERIES

**Lithium-ion batteries (Li-ion)** are the latest available technology. There are various styles of lithium-ion batteries, some of which have a greater tendency to burst into flames. You have probably seen the news stories of hover-boards, cell phones, and even airliners having fires caused by Li-ion batteries.

The version showing the most promise for off-grid battery bank use is the LiFePO4 (lithium iron phosphate) battery. LiFePO4 batteries are generally lighter and smaller than lead acid batteries, but size is usually not a great concern for off-grid battery banks. Significantly, they are maintenance-free and have longer life spans than lead acid batteries. However, their charging requires a high level of monitoring, and few charge controllers currently on the market can properly charge them. Both over- and undercharging can permanently damage these batteries. The company Tesla has produced a promising lithium-based battery system called Powerwall that may be a great player in the battery market.

# WEIGHING THE OPTIONS

We recommend starting with flooded batteries for your first battery bank because they are cheaper to buy, cheaper to replace, longer lasting, and take abuse better. While the maintenance-free batteries require less hands-on work than the flooded batteries, they are also less equipped to handle mistakes. For example, if a flooded battery is overcharged, water that boils off can be replaced. Overcharge a gel cell or AGM battery and you have to replace the full battery. Keep your batteries clean, ventilated, and away from flames or sparks. They prefer a temperature between 32 and 80°F.

Most of your choices will be 6-volt batteries, which you will connect together to make 12-volt banks. Only use deep cycle batteries; do not use car batteries. GC2 golf cart batteries are a good economical choice, pricing in at $85 for 208 amp-hours. Larger L16-sized flooded batteries that are typically used in floor-scrubbers and similar tools are also a good, if heavier, choice.

Hopefully, by the time you have to replace your batteries, we will have some newer and better choices, perhaps in lithium-based batteries or even fuel cells.

# CHAPTER 10
# CARE AND FEEDING

## Making Your Batteries Last

Batteries are important to an alternative power plan and you need to take good care of them. They have a limited life span that you want to extend as much as you can; you certainly don't want to shorten their life span. So let's look at how to best care for our batteries. Based on what we discussed in Chapter 9, we will focus on caring for flooded lead acid batteries, also known as wet cells.

# FLOODED CELL MONITORING

As flooded batteries are charged, some of the liquid boils off. The top of the lead plates must never be exposed, so it is important to monitor the fluid level of each cell by removing each cap and peering inside. *Always* remember to use safety glasses when checking inside of a battery.

Check your levels at least once a month. If levels are low, top off each cell using *only* distilled water; this is known as watering. *Never* add acid, tap water, or bottled water to a battery. *Only* check and water a battery after a full charge.

# WATER, PURE WATER

In a typical, healthy, fully charged FLA battery, you should have a mixture of 65 percent water and 35 percent sulfuric acid. While you should never add acid to your batteries, you should add water, probably on a monthly basis. A small amount of water is lost during each charging process and must be replaced. It is important to use only pure, distilled water; any impurities can interfere with the chemical reaction and shorten the life of the battery.

Not all distilled water is of the same quality. We recently tested two different sources of commercially sold distilled water and measured each with a parts-per-million (PPM) meter. The first sample measured zero contamination, as we expected. But the second sample measured 17 parts per million of some type of contamination.

Distilled waters of different qualities

This being a book on self-reliance and being prepared, we do not want to put ourselves in the position of losing our alternative power simply because we ran out of distilled water. Buying commercially sold distilled water is a cheap enough purchase to stock up, but eventually, you will run out. A much better solution is to invest in a water still so that we can produce our own distilled water. And it is not a big investment either.

We purchased the 2 Quart Per Day Prime Water Universal Water Purifier, which is simplicity itself and can be found on Amazon. We went with the smaller 2 quart per day water still as our needs are not great. We use about a gallon a month to water the battery bank.

The Universal Water Purifier consists of three parts; the raw water pot, the floating collection bowl, and the condensation lid. Here's how it works.

The three parts of the Universal Water Purifier

1 Fill the main pot with a few inches of water and float the collection bowl on top of that water. You want just enough water in the pot so that when the collection bowl is full, it will rest on the bottom of the pot without sinking below the water level.

2 Place the condensation lid on top of the pot. Cover the lid with cold water.

The condensation lid rests on top of the pot

3 Bring the water in the main pot up to a simmer (not to a boil, as you don't want it to splash). The steam will rise to the bottom of the condensation lid, where it will cool, condense, and drip into the collection bowl.

4 You will periodically have to replace the cold water that covers the lid as it heats up. As you do that, empty the

The collection bowl has gathered distilled water

collection bowl into clean jug or jar where you will keep your distilled water.

# A GLASS HALF FULL: STATE OF CHARGE

Every battery has a **rated capacity**, which is the expected energy capacity of the battery. Once you start using your batteries, your goal is to know what your battery's state of charge is. The **state of charge (SOC)** indicates how much of the battery's capacity is still available. The easiest but least accurate method for measuring SOC is to use your handy $5.00 voltmeter and measure the battery bank's voltage. For the 12-volt bank being used in this book, a 50 percent SOC is 11.9 volts. Once you reach that voltage, we recommend shutting down the power loads connected to your system to prevent damaging your batteries.

You may be wondering why a 50 percent state of discharge on a 12-volt battery is 11.9 volts, and not 6 volts (0.5 * 12v = 6v). The reason comes from chemistry. When a 12-volt flooded lead acid battery reaches 10.5 volts, the chemicals in the acid that react with the plates are exhausted; there is not enough energy left to continue the chemical reaction. So 10.5 volts is at a zero SOC, not at zero volts.

Never take a 12-volt lead acid battery bank below 10.5 volts, as doing so will reduce the life span of the battery bank. Make your measurement when the battery bank is idle, not when it is charging and not when it is powering any loads. The best way to take your measurement is to open the main fuse or circuit breaker that connects the battery bank to the rest of your system and then give it a few minutes to settle down before making the voltage measurement. Power coming in from the solar array or going out to your appliances will skew the results.

Here is a table for your reference:

| SOC | VOLTAGE |
| --- | --- |
| 100 percent | 13.38v |
| 90 percent | 13.09v |
| 80 percent | 12.80v |
| 70 percent | 12.51v |
| 60 percent | 12.22v |
| 50 percent | 11.93v |
| 40 percent | 11.64v |
| 30 percent | 11.35v |
| 20 percent | 11.06v |
| 10 percent | 10.77v |
| 0 percent | 10.48v |

A more accurate way to track the SOC is by monitoring how many amps are put into the battery versus how many are taken out. The charge controller we use and recommend (MidNite Solar Classic) has a $75 accessory called the Whiz Bang Jr. (pictured on page 119) that performs exactly this function. Acting like a car's fuel gauge, it tracks how SOC increases as the batteries are charged and decreases as the batteries are drained.

A third method you can use with flooded lead acid batteries is to use a hydrometer to measure the specific gravity of the battery acid. We will explain this method in Appendix C.

# SNAKE OIL? THE DESULFATOR

As flooded lead acid batteries age, a scaly layer of sulfate crystals builds up on the plates in a process called **sulfation**. Sulfation reduces the storage capacity of the battery, increases the time it takes to recharge the battery, and reduces the battery's usable life. The sulfation process is worsened when the batteries are exposed to high temperatures, left in a discharged state, or discharged too deeply.

Most charge controllers have an **equalization** feature. This is a fancy term for overcharging a battery on purpose. Please note that equalizing is *only* acceptable for flooded lead acid batteries, *not* for gel cell or AGM batteries. Some folks recommend equalizing on a monthly basis, while some suggest every six months or even yearly. If you are going to do it at all, monthly seems the best choice. By overcharging the battery for a couple of hours, you cause a rapid boiling of the acid, which does two things: It mixes the acid, which tends to stratify over time, and it breaks off the crystals that form on the plates, thus removing the culprit behind sulfation.

The downside to equalization is that when scales are knocked off the plates they settle on the bottom of the battery case. Over time, the scales build up. When the scales reach the bottom of the plates, the plates short out and the battery dies.

Enter **desulfators**. These devices continuously pulse the battery with random frequency voltage spikes, which in theory dissolve the crystals into the acid solution. It's a great idea, but an untested one. They say desulfators can double the life of your battery, and if so they are worth the price; on batterylifesaver.com, we found one for $114. We are on the third year of using one of these, so we need four or more years to know if it is working. We could cut one of our batteries in half and look at the plates to see, but we are hesitant to do this. Good evidence is starting to show up on the Internet that they are effective.

Even with an effective desulfator, you should still equalize your battery bank at least every six months. A flooded lead acid battery that does not go through a good boil on occasion will suffer from stratification, which is where the acid settles into layers and cannot produce a steady, strong supply of power. The acid on the bottom of the battery case will be of a different concentration than the acid at the top. Boiling mixes the acid back together.

# HE'S DEAD, JIM: BATTERY LIFE AND CHARGING

The life of your batteries will be affected by how much they have been drained before recharging and how many drain/charge cycles they experience. For our purposes, we are considering a battery empty when it reaches a 50 percent SOC (11.9 volts on a 12-volt bank). In an emergency, you can go down to 20 percent (11 volts on a 12-volt bank), but realize that this will affect the overall life span of the battery bank.

Leaving batteries in a low state of charge will shorten their life span, so recharge your bank as soon as possible. A moderate daily discharge of 10 percent to 20 percent before recharging will maximize the battery's life span. On a typical day, we will let our bank's SOC drop to 85 percent at the lowest before charging back up to 100 percent. This will result in a higher number of discharge/charge cycles before battery death than if we dropped to 50 percent SOC every day before recharging.

Batteries have an **internal resistance** to taking a charge. To overcome internal resistance, some power will be lost to heat, so more power is needed to charge a battery than can actually be stored for later use. This means that if you want to charge a battery, you must input voltage at a higher level than the battery's storage capacity. For example, the 12-volt battery bank requires 14.6 volts to charge.

Batteries lose charge efficiency over time. Flooded lead acid batteries start out at about 95 percent efficiency, meaning only 5 percent of your power is lost to heat. As they age, the efficiency drops to around 85 percent. The GC2 lead acid golf cart batteries we use have a life span of two to seven years if they are properly cared for. Hopefully they will last longer with the desulfator.

You might be tempted to store up some batteries for a stormy day. This would be fine if you could buy batteries and battery acid separately and only mix them when you wanted to put them into service. Unfortunately, you cannot buy them that way; at least, we have not found anywhere to buy dry batteries. A battery starts its chemical process the moment the acid and plates meet each other and setting them on a shelf—yes, including the store shelf—starts the clock on their life span. So as soon as you buy them, use them.

# CHAPTER 11

# CHARGE IT!

## Using a Charge Controller

What should you do if your solar panel or solar array outputs a slightly higher voltage than that of your battery bank?

You could connect the panel or array directly to the battery bank to charge the batteries. For example, many smaller panels in the 15- to 100-watt range put out 18 volts, which can charge a 12-volt battery bank. But many of the larger panels (over 200 watts) put out more than 30 volts, which is too high of a voltage to charge a battery bank directly without damaging the battery.

Voltage problems aside, batteries need to be charged in stages. Otherwise, they may be damaged and not fully charge. To prevent this damage, you really need a charge controller.

# THE ROLES OF THE CHARGE CONTROLLER

Batteries must not be charged with too much current or voltage and must not be overcharged. The **charge controller** protects your batteries from damage caused by any of these elements during charging. The charge controller is also responsible for charging the batteries in the most efficient manner, wasting as little power as possible. And some charge controllers also protect the batteries from being too deeply discharged by automatically disconnecting loads when the bank's voltage gets too low.

## CHARGING SPECIFICATIONS

To properly use your charge controller and charge your batteries, you will need the battery manufacturer's charging specifications. It's important to look at these specifications because not all batteries are charged in the same manner. The voltages and sometimes the currents for each stage of a charge can vary. Sometimes the charging information can be difficult to find; for example, not all batteries are actually manufactured by the company that sells them. The Energizer GC2 golf cart batteries we use actually appear to be manufactured by U.S. Battery (www.usbattery.com, model 1800 XC2). It took us a few days of Internet searching to find the specifications for these batteries.

## TEMPERATURE COMPENSATION

The voltage necessary to charge a battery varies depending on the temperature of the battery. Batteries convert a chemical reaction into electricity, and as in most chemical reactions, they are affected by temperature. Heat accelerates the reaction and cold slows it. Therefore, as the battery temperature drops, the charging voltage must slightly increase to compensate. A good charge controller will include a temperature sensor that you can stick to the side of one of your batteries. The manufacturer's specifications will tell you what compensation voltage to enter into the charge controller.

# CHARGING YOUR BATTERY

## BULK STAGE

At the beginning of a charge cycle, you will be dumping all the power you can into the battery. This is called the **bulk stage**. The battery's charging specifications will tell you the maximum rate at which you can dump power into the battery. Many charge controllers allow you to set the maximum rate. This rating is often not a significant factor because your charge starts at the beginning of the day when the sun is low in the sky and your solar array won't be producing much power. By the time the sun is high enough in the sky, your bulk stage will probably already be over. The goal in the bulk stage is to keep dumping power into the battery until the battery voltage reaches the absorb voltage. By the time the battery reaches the end of the bulk stage, it should be around 80 percent full.

With the 208 amp-hour GC2 batteries we use, 21 amps is the maximum charging rate. By the end of the bulk stage, the battery bank will be at 15 volts.

## ABSORB STAGE

Once the battery has received enough power in the bulk stage (each battery differs, so check the manufacturer's specifications), the charging method changes. If you continued dumping in power in an unrestrained manner, the battery voltage would keep climbing, causing the acid to boil and eventually ruining the battery.

In the absorb stage, we no longer want the battery voltage to climb; rather, we want it to remain constant as the remaining 20 percent of the battery is filled. To do this, you will set the charge controller to decrease the amount of current going into the battery.

This graph shows battery voltage and charging current. The dark line on the graph represents the battery's voltage and the lighter line represents its charging current. The absorb point is when the voltage stops climbing and levels off and the current starts to

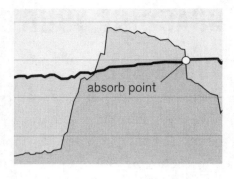

absorb point

drop. On a side note, this graph comes from our charge controller's ability to connect to the Internet to log its data.

You can determine when the absorb phase is complete by monitoring the charging amps, and when you see the charging amps level off for 15 to 30 minutes you then know the battery is fully charged. The point where the charging amps stop dropping is known as the **ending amps**. Ending amps will tell you at what charging amperage the battery is full and you can stop charging.

In a situation where you don't have heavy loads turning on and off, ending amps is a great way to end the absorb stage. But if you have heavy loads turning on and off throughout the day, those loads are very likely going to occur during the absorb stage. The extra demand of multiple loads turning on and off causes more current to flow through the charge controller than just the amount being used to charge the batteries; this fools the charge controller into thinking all of the current is being used to charge the batteries, which skews the ending amp reading.

To avoid the confusion that comes with charging batteries and multiple loads at the same time, we are left with the easier method of ending the absorb stage: time. Battery manufacturers will state how long their batteries should remain in the absorb stage, which is usually two to three hours. End time is programmable on the charge controller. When the time's up, absorb is finished.

## FLOAT STAGE

This is the final stage, and it will continue until the sun goes down and there is no longer enough sunlight to feed the battery bank. Float stage is a constant-voltage low-current stage that is just above the battery bank's resting voltage; it simply maintains the battery. In the GC2 battery bank, the float voltage is 13 volts.

## REBULKING

During the typical charging day, you will probably be using some electricity at the same time the battery is charging. If your solar array produces enough power to both charge your battery bank and to meet your daytime power use, then all is good. But if you should happen to have a need that exceeds what your array is making, then the shortfall will be pulled out of the battery. If you are in either the absorb or float stage and you draw a significant amount of power from the battery, you will probably want to re-enter the bulk stage and top the battery off again, assuming there is enough daylight left to do so.

This is called rebulking, and the charge controller has a setting for what voltage you should use to trigger rebulking. We use 12.5 volts, which is about 70 percent SOC.

## MPPT: GETTING THE MOST OUT OF THE PANELS

During the charge cycle, the power coming out of your solar array will vary. The sun will climb and drop in the sky, clouds will come and go, and the temperature of the panels will change. The solar power industry addressed this variability by inventing a charge controller method called multi-point power tracking (MPPT). This type of charge controller will pull as much power as possible from your array. Every three minutes or so, it scans the voltage coming in from the array and then modifies the voltage being sent to the battery bank to best use the available power. It is sort of black magic.

MPPT also has the ability to use the high voltage coming in from the array to charge a lower voltage battery bank. If you are using a charge controller that does not have the MPPT feature, then the solar panels you are using need to have a similar voltage to that of the battery bank. For a 12-volt battery bank, this limits you to using the smaller wattage 18-volt panels. However, due to its ability to lower the voltage it receives from the solar array, a charge controller with the MPPT feature can take the higher voltage, higher wattage panels and maximize charging the battery bank without damaging the batteries with too much voltage.

Not only can you use higher voltage panels with MPPT, but you can also connect multiple solar panels in series to save wiring and to use smaller sized cables. As a reminder, connecting your solar panels in series reduces the wiring you have to run, and allows you to use thinner cables due to the lower current. In series, the output voltage of each panel is added together (Chapter 9). Our series-wired array has produced 124 volts on occasion. Only an MPPT charge controller can handle these high voltages and charge your battery bank.

# GETTING THE POWER INTO THE BATTERIES WITH PWM

Another type of charge controller is the pulse-width modulation (PWM) controller. This type of controller lacks the MPPT's ability to work with high voltages and to extract the most power from your solar array. However, PWM adjusts the power going into the battery bank by supplying the power in variable pulses. It rapidly turns the power to the batteries on and off and varies how long the on and off times are. This is similar to how light dimmers work; they vary turning the light on and off at a rapid rate that can't be seen, and the brilliance of the light changes based on how long the "on" is compared to the "off."

Morningstar Prostar

If money really is an issue, two PWM charge controllers you should look at are the Morningstar Prostar for $180 and the MidNite Solar Brat for $125. Note that these both are smaller controllers; you will have to use smaller solar arrays with panels that have lower voltage.

MidNite Solar Brat

# SIZE MATTERS: ONLY BUY IT ONCE

We use and strongly recommend the MPPT MidNite Solar Classic charge controller from MidNite Solar, available in 150-volt, 200-volt, and 250-volt models. We recommend the 150 or the 200. The numbers reflect the maximum combined voltage from your solar array. As the maximum voltage rating increases the amount of current flowing to the battery bank decreases, which is why we don't recommend the 250 model. You should not need to feed that much voltage to the charge controller from your solar array anyway.

The MidNite Solar Classic is, in our opinion, the best charge controller on the market. It has excellent features, is well built, has wonderful support, and has a large user base and discussion forums. The MidNite Solar website also has a large library of training videos. Their sense of humor is also a joy.

Yes, there are other, lower-priced choices (the Classic 150 sells for around $700), but we promised you that we would try to help you avoid making the mistakes we made. We now own three charge controllers. As our experience over the past four years off-grid has grown, we kept replacing our charge controllers as we refined and improved our system. The result was total satisfaction with our MidNite Solar Classic.

MidNite Solar Classic 200

Appendix B lists the settings we use in our MidNite Solar Classic 200. We are not stating that the settings we use in our controller will be perfect for your situation; this is especially true for the battery settings, as you may use a different battery than we used. This appendix is just to serve as an example.

# PART THREE

# I'VE GOT THE POWER: BRINGING YOUR SYSTEM TO LIFE

# CHAPTER 12
# IT'S ALIVE!

## Putting Your Solar System Together

Now that we have covered all the parts, let's put a system together. This discussion will cover a 1,000-watt system, which should provide you with adequate power to meet the majority of your household needs. This exact system certainly worked for us. You are, of course, free to use different types or quantities of panels, and free to change around the battery bank as you wish. You may want to use AGM batteries rather than the flooded lead acid batteries we specify, just to save yourself some maintenance. This design is certainly not a rigid blueprint; it is simply what we have used ourselves and have been pleased with.

Let's start by listing the parts:

**Solar Panels.** Grape Solar 265-Watt Polycrystalline Solar Panel (4-Pack) at $1,200 OR Canadian Solar CS6P-265P, four panels at $232, totaling $928

**Charge Controller.** MidNite Solar Classic 150 at $695

**Battery Bank.** Energizer GC2 232 amp-hour 6-volt battery, four batteries at $85, totaling $340

**Inverter (optional).** Samlex PST-1000-12 Pure Sine 1000-watt 12-volt inverter at $380

You will need a few other odds and ends to hook everything together.

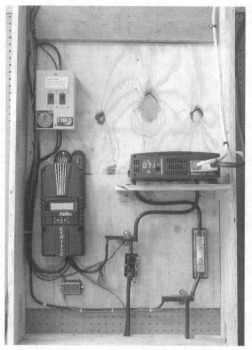

A friend's basic setup containing a solar breaker panel, charge controller, busbars, battery fuse, Whiz Bang Jr., and inverter

# WIRING

We have mentioned that the thicker the wire, the more current it can carry without loss or risk of overheating. As more current is carried through the wires, more voltage is lost in the form of heat, and that heat can cause fires. Because of this, it is important to always use wiring that is sized for the current it will carry. In fact, this is exactly how fuses work; fuses are short pieces of wire of the exact thickness needed to carry their rated current, and nothing more. When you pull too much current through the wire in the fuse, it overheats, burns up, and breaks the circuit. Here's a wiring size chart to work with.

| GAUGE | CURRENT (AMPS) |
|-------|----------------|
| 1/0 | 125 |
| 1 | 110 |
| 2 | 95 |
| 3 | 85 |
| 4 | 70 |
| 6 | 55 |
| 8 | 40 |
| 10 | 30 |
| 12 | 20 |
| 14 | 15 |

# BATTERY CABLES

To connect the GC2 batteries, the easiest solution is to pick up an assortment of golf cart battery cables. These are typically 4 gauge cables. For our battery box, we found the 7- to 9-inch cables worked well to interconnect the batteries. We bought ours from Amazon; you can also find them at golf cart stores and some farm supply stores.

To connect the battery bank to the charge controller you can use the same 4 gauge cable we just discussed, but you may wish to use 1 or 2 gauge for future use, should you ever expand your battery bank to supply greater power later. By using 1 or 2 gauge wire to connect the battery bank to the charge controller, you will avoid rewiring later. Try to keep the battery bank within four feet of the charge controller. The shorter the cable connection, the lower the voltage loss. The charge controller bases its computations on what it believes the battery bank voltage to be, and during peak charging the amperage flowing through the cables can be rather large. If the cable is on the thinner side, then there will be more voltage loss through the cables, which causes the charge controller to believe the battery voltage to be lower than it really is. A thicker, shorter cable will minimize that voltage error.

## INVERTER CABLES

The heaviest cables should be those between your battery bank and your inverter, if you install one. We mentioned earlier that we have a microwave oven that we use on very rare occasions. When in operation, our inverter pulls 110 amps from the battery bank. So we strongly suggest that you use 1/0 or 1 gauge wire for the inverter.

## SOLAR PANEL CABLES

**Warning**: Be extremely careful when wiring solar panels. Each solar panel typically outputs 30 volts of electricity. This is not a trivial amount of voltage. When wired in series, each panel's voltage is added to all the other panels, meaning four panels can produce over 100 volts! Take extreme care when doing your wiring on the panels; you may wish to cover all the panels with a blanket to block the sun while doing the wiring.

The factory-installed cables on your solar panels are typically 10 gauge. The panels we are recommending in this book can provide a maximum of 10 amps, so the 30-amp rating on 10-gauge wire is more than enough. If you don't have any shading problems and are able to connect your four panels in series as we suggest, they will still only supply a maximum of 10 amps. (Remember that in series wiring, each panel's voltage is added together but the amps remain the same). If, however, you have to connect your panels in parallel (with either all four panels in parallel or two parallel strings of two panels in series), the current will add up. If you have to run all four in parallel, then you will be facing a maximum current of 40 amps, so the cable between the combiner box where the panels are connected in parallel and your charge controller will need to be at least 6 gauge with a 55-amp capacity.

The MC4 connector cables that come with the panels are usually only around 30 inches long, so you will need an extension cable to reach your charge controller. The easiest thing to do is buy an MC4 extension cable twice the length you need and cut it in half. This will give you two cables: one with a male MC4 connector and a bare end, and the other with a female MC4 connector and a bare end. The connectors will be used to connect to the panels or combiner box, while the bare ends will be stripped and screwed down on the charge controller connections. Your charge controller has screw down connectors instead of MC4 connectors, which are used by the solar panels.

Try to keep your panels within 50 feet of the charge controller. If you cannot find an extension cable long enough, you can buy the correctly sized wire from any home improvement store or hardware store, buy the MC4 connectors from Amazon or a local store, and build your own cables. Many YouTube videos show how to install the connectors.

If you have to run your panels in parallel, you can either use a combiner box or simply use "Y" cables to connect them. For example, if you run all four in

Y adapter

parallel, you will have a total of four male and four female cables from the panel array to deal with. A 4-Y cable has four connectors on one end and a single connector of the opposite gender on the other. Plug the four-panel male connectors into one of the Y adapters, and then plug the four female panel connectors into the other Y adapter, and you are then left with a single male and a single female connector. Those will each connect to a 6-gauge (or thicker) extension cable and run to your charge controller.

We recommend that you purchase an MC4 assembly/ disassembly tool, which is a type of spanner. Once you connect a male and female MC4 connector, they are extremely difficult to disconnect without the tool if you ever need to make some changes.

MC4 assembly/disassembly tool

## COMBINER BOX

If you can connect your panels in series and you only have a single male and single female cable going to your charge controller, no Y adapters are needed. It will be a simple installation using standard MC4 10 gauge cables the whole way.

But if you have any parallel connections, then you may wish to use a combiner box, which is typically located out at the panels. A **combiner box** allows you to feed each parallel connection into a weather-protected box. Each

Highlighted area shows combiner box location

parallel connection is run through a fuse or DC circuit breaker, and then they are connected on a busbar rather than with a Y adapter. This arrangement does two things: It protects each parallel connection from any over-current damage, and it gives you a convenient way to turn panels off when testing.

# LET'S GET WIRED

Your solar panels are hung on your racking, so let's hook them together. If you are using an MPPT charge controller, hook all your panels in series if you do not have any partial shading from trees or structures to deal with. In order to make these series connections, number your panels 1, 2, 3, and 4. Then follow the steps below.

Each panel has two wires, one with a male MC4 and one with a female MC4. Connect the male cable on panel 1 to the female cable on panel 2. Next, connect the male cable on panel 2 to the female cable on panel 3; then, connect the male cable on panel 3 to the female cable on panel 4. This should leave you with a female cable on panel 1 and a male cable on panel 4. These now need to be connected to your charge controller.

Earlier on page 110, we suggested buying an extension cable twice as long as necessary and cutting it down to two cables with bare ends. We also suggest that you not strip the bare ends until you will be using the cables. Just cut the cable in half and leave the ends without connectors alone. This way, you can run all your cable without worrying that the two ends might touch and make a short circuit. The risk can also be minimized by covering your array with a blanket during wiring.

Once the cables have been run to the charge controller, go ahead and strip a half-inch of insulation from the end of each cable. This is one of the places to be careful not to get shocked. Cover the panels if you have not already done so. If you used a combiner box, turn off the circuit breakers or remove the fuses.

Once you have stripped the wires, you need to identify positive and negative. If you have the panels covered with a blanket, go uncover them. If you used a combiner box with fuses or circuit breakers, make sure the fuses are installed or the breakers are on.

Set your voltmeter to at least 200 volts DC and connect a meter probe to each of the two cables coming in from the panels. Use alligator clips to avoid touching the bare wires with your bare hands.

The value displayed on your meter is not important right now, but look and see if a "minus" symbol is being displayed. If there is, you have the probes from the meter reversed: The black probe is connected to the positive cable, and the red probe is connected to the negative cable. If there is no "minus" displayed, then the red probe is connected to the positive cable and the black to the negative cable, which is what you want.

Now that you know which cable is positive and which is negative, tag them in some manner to help you remember, perhaps by attaching a piece of masking tape to the positive cable. Then, just for sanity's sake, make sure you have the probes correctly connected to your meter. The black probe should be connected to the COM, or "common," socket.

## SAFETY TIP

When you bring the cables in from the solar panels, you need to run the positive cable through a fuse or DC circuit breaker to protect the circuit from an overload that could overheat and cause a fire. We use ANL-style fuses, but the choice is up to you. Another good and inexpensive choice is the automotive blade-style fuse block and fuse, called an ATC-style fuse. Determine the maximum current that might come in from your solar array and install a slightly larger fuse/breaker. If your four panels are in series, you should use a fuse/breaker a bit larger than 10 amps. If all four are in parallel, you should use one a bit larger than 40 amps.

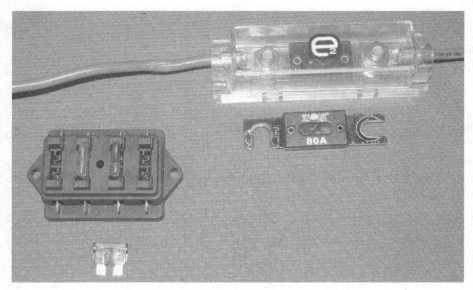

ATC fuse and block (left) and ANL fuse and block (right)

# CONNECTING THE CHARGE CONTROLLER

Start by reading the manual and watching the videos that may come with your charge controller. MidNite Solar includes a CD with instructional videos, and their website has even more information. Get to know your controller before starting to wire it.

The MidNite Solar Classic is a very advanced controller. While we quickly got ours up and running, it took a lot longer to play around with it and review the results before we felt we knew it well.

When you are ready, follow the directions to remove the faceplate to access the screw terminals. It is a good idea to open the fuse/breaker to make sure no power goes into the controller from the solar array. Feed the cable you tagged as the positive cable into the controller housing and fasten it to the positive screw terminal that the manufacturer has labeled for the solar panel.

Once you've connected the positive cable to the correct terminal, do the same for the negative. Leave the panel's fuse/breaker off until you are all done. Be careful not to mix wires and accidentally cross the positive and the negative cables. Also, do not accidentally connect your panels to the terminals on the controller labeled "Battery."

# BATTERY BANK

Earlier we discussed interconnecting your batteries. Now you'll learn how to perform that task. Get your eye protection for this part!

As mentioned in Chapter 9, you can increase the capacity of a battery bank while keeping the voltage at 12 volts by adding a second string of batteries in parallel with the first string. But remember that you really should not go over two parallel strings, and you really, really, *really* don't want to go over three strings. If two parallel strings of batteries are not enough, then you will either need to use larger capacity batteries (our recommended GC2 batteries have 208 amp-hours, so two strings in parallel provide 416 amp-hours), or you need to use a 24- or 48-volt battery bank.

Before you start, use some petroleum jelly to lightly coat all the battery terminals and all the ends of your battery cables. This will help prevent corrosion from building up on the terminals and cable connectors.

The goal is to connect the batteries in a manner that charges and discharges them all in as balanced a manner as possible.

Label your batteries as 1, 2, 3, and 4. We will start with series connections, which will result in pairs of 6-volt batteries that will then provide you with 12 volts. This is an easy step. Using one of your golf cart battery cables, connect the negative terminal on battery 1 to the positive terminal of battery 2. You will be left with an empty positive terminal on battery 1 and an empty negative terminal on battery 2. Use your voltmeter to measure the voltage across those two empty terminals and you will read somewhere between 11 to 13 volts, depending on how much the batteries have been charged.

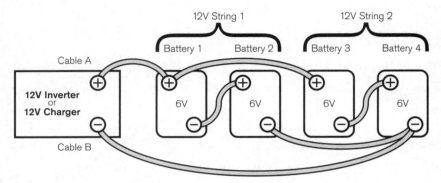

Battery bank wiring example

Using a second battery cable, connect the negative on battery 3 to positive on battery 4. Use your meter to check the voltage on the empty positive on battery 3 and the empty negative on battery 4. You now have two independent series strings of two batteries each. Remember that batteries in series add their voltages together but not their amperages.

Now you'll want to connect the two strings together in parallel to double the amp-hour capacity. Take caution with this step, because batteries can supply an incredible amount of current. You never want to short-circuit your batteries. The resulting sparks, fire, and possible battery case explosion can result in death or injury.

With this in mind, you can connect the two strings in parallel. You will be using two of your battery cables. Connect the first cable between the empty positive terminal on battery 1 and the empty positive terminal on battery 3. Visually double-check the cables you have run so far, according to the instructions above. If the cables are connected correctly, you can use your last cable to connect the empty negative terminal on battery 2 to the empty negative terminal on battery 4. Again, double-check all your cables by closely looking them over and comparing your wiring to the instructions we have given.

If everything is in order, you will have built a battery bank!

# CONNECTING THE BANK TO THE CONTROLLER

Hopefully, you were able to find a couple of golf cart battery cables of at least 4 gauge that are long enough to reach between your battery bank and your charge controller. We recommended you maintain a distance of four feet or less between the bank and the controller to avoid excessive voltage loss. If you could not find any pre-assembled cables consisting of a wire with terminal lugs on each end, you can buy the correct sized cable at most home improvement or hardware stores and construct the cable that you will need. The best place to find the properly sized eyehole-style connectors that you will need for your cables is in the electrical aisle or the battery cable aisle of automotive parts stores. You might also find an electrical supply business that will make your cables. It will take considerable heat to solder on those connectors.

To help balance the batteries in the bank, connect the positive cable going to the charge controller to the positive terminal on battery 1, and connect the negative cable to the negative terminal on the battery 4. Once you've done this, the positive charge will be on the first battery of string 1 and the negative charge will be on the last battery of string 2. Together with the cables you already installed that tie the two strings together in parallel on those terminals, you will have two additional cables on each terminal that will connect to the charge controller.

The positive cable running from the battery bank to the charge controller needs to be protected with either a fuse or DC circuit breaker. Your 4 gauge cable can handle 70 amps, so you should use a fuse/breaker a bit smaller than 70 amps; if there is an overload, we want the fuse/breaker blowing before the cable burns up. Those batteries can supply well over 70 amps if shorted, which will burn up those heavy cables if they are not fuse or breaker protected. Leave this fuse/breaker off for now.

Once you pass the positive cable from your battery bank through the fuse/breaker, you will need to connect the positive to a busbar. Next, connect the negative from the battery bank to its own separate busbar. You are using these busbars rather than directly connecting the battery bank cables to the charge controller because the busbar creates an area for you to tap into the positive and negative currents you will need to run your household devices.

Busbar

## USING THE WHIZ BANG JR.

MidNite Solar has a good sense of humor, and one place that it shows is in the naming of the Whiz Bang Jr, an optional accessory for their charge controllers. The Whiz Bang Jr. will always let you know how much capacity is left in your battery bank at any given time by measuring the bank's state of charge. Every amp of power that flows in or out of the battery bank will go through the Whiz Bang Jr. In this way, the Whiz Bang Jr. acts like the gas gauge in your car: Put gas in the tank at the gas station and the gauge goes up; drive around and it goes down.

The Whiz Bang Jr. is optional but recommended. If you buy one, now is the time to install it. This device is installed between the negative cable that goes to the charge controller and the negative busbar we just mentioned. Note that if you intend to use the Whiz Bang Jr., you will need to purchase an additional shunt supplied by MidNite Solar to measure the current. Follow the manufacturer's instructions for installation.

Using a wire of the same gauge as before, connect the positive battery busbar to the battery positive screw down terminal on the charge controller, and the negative battery busbar (or Whiz Bang Jr. shunt) to the negative screw down terminal on the charge controller.

Whiz Bang Jr.

# VENTILATION

Before you power on your charge controller to start your first charging, you need to make sure your ventilation system is working. If you spent the extra for AGM batteries, then just having some air holes in the top of your battery box will be sufficient. However, you'll need to do more work if you are using flooded lead acid batteries.

On page 84, we discussed and showed an example of a ventilation system. This is critical to have up and running when charging your batteries—or in other words, *always*. You must use an induction motor fan like a PC cooling fan. Any regular fan motor has brushes that spark and could detonate the very hydrogen gas you are trying to ventilate. Run the fan's power wires, which can be low-gauge hook-up wire, up to the positive and negative busbars. Make sure you pay attention to which wire is positive and which is negative, as inductive motor cooling fans are polarity sensitive. Verify that the fan is running and pulling air from outside the box into the box. Also verify that you have a ventilation tube near the top of your battery box that reaches outside air, then close up the battery box. If the box is not closed, the fan will not blow the gases out through the ventilation tube.

# POWER ON!

Let's review the startup instructions for your charge controller again, just so you will be ready and know what to expect. When you are ready, turn the battery bank breaker on or install your fuse. The charge controller should come to life and you will have to make a few choices on its settings panel. This will be explained in the instructions that come with the controller.

Next, you can go ahead and turn on the solar array breakers or install your fuses. If your solar panels are still covered with a blanket, go ahead and uncover them.

If using the MidNite Solar Classic, on the charge controller's display panel, press the "Menu" button once and then the "Status" button. You will be looking at the first of three status displays, or four if you have a Whiz Bang Jr. installed. You can push the "Status" button to move through the other displays. You should be seeing current flowing to your battery bank.

# CONGRATULATIONS, IT'S ALIVE!

You've put your system together and got it up and running. Now it is time to do something with it.

In the next two chapters, we will be looking at two different approaches to running your home's devices from your solar power system.

# CHAPTER 13
# LESS IS MORE, MORE OR LESS

## The Merits of DC Power

Unless you live on a boat, in an RV, or in a camper, it is a safe bet to say that your electrical appliances, lights, and outlets are all operating on 120 volts of alternating current. In 1886, George Westinghouse started building electric power plants based on Nikola Tesla's AC designs. But before that point, electrical service was based on Thomas Edison's direct current designs.

In this chapter, we are going to turn back the hands of time and convince you to adopt Edison's DC system.

## THE WAR OF THE CURRENTS

Edison is famous for his adage that "genius is 1 percent inspiration, 99 percent perspiration." It is obvious that he dearly held to this statement, as he relied heavily on experimentation for his discoveries instead of formal education, of which he had little to fall back upon. Edison was only able to invent by trial and error, rather than by design, science, and mathematics.

Nikola Tesla provided a stark contrast to Edison's methods. Tesla received years of formal training at engineering schools in Graz,

Austria, and Prague, Czech Republic. Tesla earned the equivalent of bachelor's degrees in physics, mathematics, mechanical engineering, and electrical engineering from Austrian Polytechnic Institute by passing the exams. However, it was Tesla's genius more than his formal education that brought about his discoveries and inventions. Tesla was known to suddenly stop and whip out paper and pencil to sketch out fully formed inventions that just appeared in his mind. He reported that he could think about an idea and see it fully modeled and functioning in his head. He could solve advanced mathematical problems on the "blackboard in his mind." Tesla could be regarded as a creative savant.

It is between these two differing personalities that the War of the Currents was waged. A sudden need for the distribution of electrical power was realized with the invention of the incandescent light bulb. As the inventor of said light bulb, Edison was given the responsibility of discovering how to deliver electricity to light numerous bulbs across the city. This task proved to be much more daunting than the creation of the light bulb itself.

Look closely at a light bulb. You'll see that it consists simply of a strand called a filament that heats up to a high temperature when electricity is applied. Electricity causes a filament to glow brilliantly without being consumed in the heat. Edison's "genius" was nothing more than his trying dozens of materials to act as a filament before he found one that did not immediately burn up in the vacuum of his bulb. This trial and error method would not work as well for building power plants and a distribution network. Tesla, the day after Edison's death, summed up Edison's brute force approach to inventing this way:

> *If he had a needle to find in a haystack he would not stop to reason where it was most likely to be, but would proceed at once, with the feverish diligence of a bee, to examine straw*

*after straw until he found the object of his search. ... I was almost a sorry witness of such doings, knowing that a little theory and calculation would have saved him ninety per cent of his labor.*

—*New York Times*, October 19, 1931

Edison understood direct current, batteries, and spinning a magnet inside a coil of wire to convert mechanical energy into electrical energy. But Edison could not understand the math and science behind alternators, transformers, phase relationships, and other aspects of alternating currents. These concepts were the world of Tesla, who could envision them in his head before ever building a new device. Tesla's AC induction motor is considered one of the ten most important inventions of all time, and the entire design popped into his mind in 1882 while walking in a park in Budapest.

Edison feared the loss of his energy empire if Tesla's alternating current was adopted over his own direct current. He could not adopt AC simply because he could not understand it. Edison also could not adopt a technology that had another inventor's name on it; this is evidenced by the fact that Tesla used to be an employee of Edison's laboratory, but his ideas were dismissed.

The fact that today AC power is used almost exclusively reveals that Tesla won the war. But why? What is superior about AC that brought about its acceptance? While there are many advantages of AC, the one we are focusing on here is how easily voltage can be changed by using an invention called a transformer. Okay, big deal, so what?

As we've mentioned before, wattage is equal to volts multiplied by current, and the higher the current flowing through a wire, the greater the voltage loss to heat. Higher current requires a much thicker wire to avoid substantial voltage loss to heating. And thicker wire has a higher cost and is harder to work with.

In an attempt to reduce voltage loss, Edison had to place his power plants no more than a mile apart. This proximity issue would only get worse as demand for current over those wires grew. During that time, many homes that desired electric lighting actually had a power plant built on their property, which often came with an employee to run it.

Compare this system to AC power, where a transformer could take the 110 volts that Edison's light bulbs ran on and increase it to thousands of volts. The same amount of power could be delivered over a greater distance using less current and much thinner wires. Then at the customer's location, another transformer would reduce the voltage back down to 110 volts.

A 60-watt light bulb at 110 volts requires 0.5 amps. Delivery of those 60 watts at 1,000 volts requires 0.06 amps. Multiply the energy required by one light bulb by hundreds and you can see how the wire savings adds up. Moreover, the 1-mile limit of a DC system is gone. On November 15, 1896, Buffalo, New York, received 11,000 volts of electrical power from Niagara Falls, a distance of 20 miles, over the Tesla-Westinghouse AC design.

So if AC is so great, then why are we recommending that you, the reader, use direct current instead for your alternative power system?

# LIVING LIFE 12 VOLTS AT A TIME

There are two reasons to switch from AC to DC: efficiency and distance. In the 12-volt system you are reading about in this book, the distances involved are short. We suggest you try to keep your solar panels within 50 feet of your charge controller. We also suggest that your battery bank sit less than 10 feet from your controller. Less is even better: Ours is within four feet of the controller. The largest currents in your system will be between your charge controller (or inverter, if you choose to install one) and your battery bank, so that is where your largest voltage loss will be. This can be kept to an acceptable amount by using short, heavy cables of 1/0 or 1 gauge. The second largest current that can result in voltage loss will be between your solar array and your charge controller. In Chapter 8, we recommended wiring your solar array in series as long as you don't have a problem with partial shading and you are using an MPPT-style charge controller that can accept the higher voltage resulting from a series array. By using the higher voltages from a series array, you can deliver the same power at a lower current, removing the need for short, heavy-gauge cables.

The short distances between the component parts of your system allow you to use a direct current system. But the efficiency of a direct current system really makes the argument for using direct current. (Plus, wiring codes for low-voltage DC are usually much simpler than those for AC!)

You have learned in previous chapters that your solar array gives you direct current, and your charge controller puts direct current energy into your direct current battery bank. The design we talk about in this book uses 12 volts of direct current in every aspect of your system.

So, if you want to use your existing appliances that run on the grid at 120 volts of AC, how can you run them on your 12-volt direct current alternative energy system? One answer is a device called an inverter, which we'll discuss more in Chapter 15. The problem with this option is that you will lose up to 20 percent of your harvested solar electricity in the process of the conversion to 120 volts AC from your 12 volts DC. That is a fair chunk of power to lose.

# SKIP THE EXTRA WORK

Let's take a look at how 120 volts AC is used in your home. 120 volts AC will rarely be used by your appliances in its raw form. It is only used that way in on-grid homes because that is how the electric company delivers it to you.

How many of your electronic devices use a power pack like the common wall wart? Ever looked at one closely? Let's do that now.

Look at the line that starts with "output." See the "12V"? That means that this power pack takes the 120 volts AC from the wall and converts it to 12 volts for the device it powers. Notice that little symbol next to the 12V with the  solid line over the dashed line. That symbol means the 12 volts is *direct* current, not alternating. And finally, the "1.0A" means it can provide a maximum of 1.0 amps.

So this means that you are starting out with 12 volts DC in your system, using an inverter (which loses 15 percent to 20 percent of your power) to give you 120 volts AC, and then you're turning around to use a wall wart power pack that changes your power back to exactly what you started with, 12 volts DC. And that power pack varies in efficiency from 50 percent to 85 percent. So you are losing power both to convert the power you harvest *and* to convert right back to where you started.

The approach we suggest instead sends 12 volts direct current through the house and skips the conversions.

You will find that not all devices that use wall warts or power packs run on 12 volts. For example, the power packs that our laptops use output 18 volts. But there is a solution to this apparent problem. We found a company that sells variable voltage power packs called **boost converters** that run off 12 volts DC and can convert 12 volts DC to a higher voltage. These devices are powered by a typical 12-volt cigarette plug power cable. Each power pack has a slide switch that allows you to pick the output voltage that you require.

Boost converter

Buck converter

What if your device requires a voltage lower than 12 volts? There are devices called **buck converters** that are similar to boost converters, but they have a slide switch that allows you to select a voltage lower than 12 volts. People often use these in their cars to power electrical equipment. We've used a few of these, and we've found them at automotive parts stores, automotive departments at big box stores, or electrical parts stores like Radio Shack.

# STARTING FROM SCRATCH

If you are in a remote or very rural location, you may not even have the option of grid power and we consider you fortunate. Perhaps you are building your home and do not have any pre-existing wiring. The rural property we bought when we were ready to go off-grid had a house that had been vandalized, and all the wiring had been stripped out for the copper. Scenarios like these provide you with an off-grid advantage in that you can install a 12-volt direct current electrical system without having any existing grid connected concerns.

## WIRELESS CONNECTION

No, we don't mean wireless power distribution, although that would really be nice. We are referring to a house that does not have any pre-existing wiring at all; one that will require you to run wiring to the outlets and fixtures you want.

The most common style of wiring available is called Romex. It is often used for on-grid 120-volt wiring. Romex is a flat cable containing a black wire, a white wire, and a bare copper wire, known as the hot wire, the neutral wire, and the ground wire, respectively.

When working with low voltage, you have another option that is easier to work with. It is known under a few names: zip-cord, lamp cord, or landscaping wiring. Back on page 108, we gave you a table showing what the current limit is for various sizes of wiring. In the picture to the right, you will see that the gauge of that spool of lamp wire is 16, so it can carry 10 amps.

Zip-cord

When you are pulling your wiring, going from one connection to another, each wire path you make is called a **circuit**, and each circuit will connect to your solar electric system through a fuse or DC circuit breaker. Each circuit will probably connect multiple devices, not just one. For example, if you are wiring a bedroom, a single circuit may be used for a ceiling light fixture and three or four outlets.

Before you make your wire runs, decide on the maximum load for each circuit. Think carefully about what lights and devices will be on at the same time. The total current that might be in use on a single circuit will tell you what gauge wire and what size fuse or breaker you will need. The fuse/breaker installed on each circuit must never have a larger capacity than the wiring connected to it. For example, a 16-gauge wire can carry up to 10 amps. If you were to install a 15-amp fuse or breaker and your system pulled too much current, the wiring would heat up and burn long before the fuse/breaker could protect it. Both Romex and lamp cord are available in multiple gauges, so choose wisely.

You may have noticed that Romex has three wires while lamp cord only has two. The extra wire in Romex is the bare copper "ground" wire. In grid power, 120 volts AC flows between the black "hot" wire and the physical earth under your feet, which can occasionally electrocute people. The bare copper Romex ground wire is designed to protect people from electrocution. However, your 12-volt DC solar electric system has no such connection to ground, nor does it have enough voltage to electrocute anyone through a ground connection.

## REUSING EXISTING WIRING

If you already have wiring in your house from a prior grid connection, you can reuse that wiring.

The first step is to make certain that the grid is no longer connected to your house. Verify that power is off at the utility pole, and better yet, make certain that your prior utility company has removed your power meter from the pole. Use your voltmeter on its 120-volt or higher AC setting and check a few outlets to make sure they are dead. Flip a few light switches; make sure no lights come on.

Your fuse/breaker panel will have two thick black cables coming in from the utility pole, where they are probably connected to a dual main circuit breaker or fuse. Disconnect these two cables and pull them out of the panel.

Run wiring from your solar electric system's positive busbar to one of the two terminals that you just removed the grid cables from. It does not matter which of the two terminals you connect to, as long as your cable is of a large enough gauge to run all the loads you might ever have on at the same time. It's a good idea to use the same gauge you used when you ran cable from the battery bank to the busbar.

Now your negative busbar needs to be connected to the neutral busbar in the service panel. This busbar will have many white wires and bare copper wires already connected. Using the same gauge wire you have been using to make connections to the positive busbar, connect the negative busbar that attaches to your battery bank to the negative busbar on your fuse or breaker.

If you have purchased and installed the Whiz Bang Jr. accessory, the negative wire going to the service panel needs to connect to the load side of the Whiz Bang Jr. The negative side of your power will flow from the battery bank, through the Whiz Bang Jr. shunt, to the service panel.

Once you've completed this step, you will have connected your 12-volt power to one half of your house, so the next step is to get 12 volts to the other half. Those two heavy cables from the utility pole each carried 120 volts, and your service panel has two sides, one for each cable. The power flowing from those two cables alternates between every other fuse/breaker rather than between left and right.

To get power into the rest of the house, you will need to run a short wire to jump, or connect, between the two sides of the circuit panel. Using the same gauge wire that you just ran from the positive busbar to one of the main fuse/breakers, connect the two fuses/breakers together. You will end up with one of the main fuses/breakers having two cables (the cable going to the positive busbar of your solar electric system, and the short jumper cable), and the second main fuse/breaker having just the short jumper wire connected to it.

# NEW WIRING

If your house had no pre-existing wiring and you wired it yourself as we described on page 129, you will have ended up with a bunch of wire pairs at your central equipment point where your positive and negative busbars, along with your charge controller, are located. Each one of those wire pairs represents one circuit in your house.

You should not connect these wires directly to the busbars without some form of protection. The goal is to prevent the wires in your circuit runs from overheating and causing a fire, especially inside the walls where you can't reach them. Remember that only the positive wires go through the fuse/breakers.

By this point, you have selected a wire of the appropriate gauge for handling the maximum load that your lights and devices might draw from each circuit. Remember that your fuse/breaker for each circuit must be of slightly less capacity than your wire gauge because you want the fuse/breaker to blow before the wire does. For economic reasons, we chose to use ATC-style automotive blade fuses. We purchased a marine fuse panel.

# AC VERSUS DC CIRCUIT BREAKERS

If you brush two wires together that are connected to a power source, you will see sparks. Unless the voltage is extremely high, there will not be any sparks until they touch. This is because air is an insulator, and it is hard for electricity to cross an air gap. But once the spark (called an arc) happens, the air around the arc becomes ionized. It is easy for electricity to cross the ionized air gap because it is no longer a good insulator.

You may recall that when using DC power, voltage is always constant with little, if any, variation. AC power, however, crosses zero voltage every 30th of a second as it alternates polarity. This means that it is very difficult to sustain an arc with AC. As you try to move the wires apart, the voltage reaches zero within a 30th of a second, the arc stops, and the ionization is gone.

This is where circuit breakers come in. Unlike a fuse, which has a metallic link that melts when blown or burns through to leave a large gap, a circuit breaker is more like a light switch. When the current reaches the circuit breaker's maximum capacity, a spring-trigger throws the circuit breaker "switch" open. However, when using a conventional AC breaker with DC power, there are two potential problems.

The first problem stems from the fact that the gap between the sides of the "switch" is not very large. If using DC, an arc is started when the switch opens that can continue even when the contacts are fully open. Second, the spring is not very strong, so the contacts separate slowly, helping any DC arc to continue forming. If the arc is not stopped, it will cause a fire.

DC breakers are designed with this problem in mind. The spring is stronger to move the contacts apart faster and the open gap between contacts is larger. Also, the MidNite Solar Classic charge controllers have a built-in feature called "Arc Protection." If you choose to use circuit breakers, make sure you understand and use this feature. The charge controller has the ability to recognize when an arc is occurring and will shut itself down. This cuts off the power coming in from the solar panels but does nothing for disabling the battery bank, so DC circuit breakers are still necessary.

# WIRING FIXTURES AND SOCKETS

We mentioned that 120-volt AC power uses a hot wire and a neutral wire. In actual use, it does not matter how an AC device is connected. Because AC power switches polarity 60 times a second, either wire on the device can connect to either wire in the outlet.

DC power, on the other hand, uses a positive wire and a negative wire, and the polarity *does* matter. Reverse the connection and you will probably destroy your device. This means that when you do your wiring and connect your outlets and lighting fixtures, you must pay attention to polarity. If you are using zip cord, you will see that one of the two wires will have a stripe. We suggest using the striped wire for your positive connection. Whichever wire you use, you must remain consistent on all wire runs.

Modern electrical outlets use what are called polarized plugs. In the picture here, note that the top slot is wider than the bottom slot. The male plug on your device's power cord will likewise have a wide prong and a  narrow prong. This arrangement allows the plug to be inserted into the outlet only one way: wide to wide and narrow to narrow. It is impossible to plug them together the other way. This is an important feature for preventing us from plugging in our devices backward and destroying them.

In AC wiring, the hot wire is connected to the narrow prong and the neutral wire is connected to the wide prong. We recommend following a similar plan and wiring your positive power to the narrow prong and your negative power to the wide prong. If you are using Romex for your 12-volt wiring, follow this convention by using the black wire as the positive and the white as the negative. Yes, we know that black is traditionally the negative in DC wiring, but we still suggest you use the expected color-coding when dealing with outlets and fixtures. That means you will use the black to be the positive and connect to the narrow prong.

Light fixtures have two parts. Unscrew the bulb and look inside the socket. You will find the threaded round "sleeve," and centered in the bottom of the socket you will find a round "button." In AC, the sleeve is connected to the neutral wire and the button is connected to the hot wire. For DC, connect the sleeve to negative and the button to positive, again using black for positive.

Once you have your outlets and fixtures wired and the power on, you should use your voltmeter to check the polarity of every outlet and every light fixture before plugging anything in. Set your voltmeter to a level above 12 volts DC and place the red (+) probe into the narrow slot on the outlet and the black (–) probe into the wide slot. You should see close to 12 volts with no minus (–) sign in front of the number. If you see a minus sign in front of the voltage reading, and you have the probes inserted in the correct slots, then you have wired the outlet backward and need to correct the wiring.

Now do the same with your light fixtures. Place the black probe against the sleeve and the red probe against the button at the bottom of the socket and verify that there is no minus sign displayed. If there is, fix the wiring to the socket. If there is no voltage reading at all, make sure the light is turned on, either at the wall switch or on the socket itself.

We should say more about wall switches. If you have ceiling lighting fixtures, you probably have, or desire, a wall switch to turn them on and off. Feel free to use a standard wall switch. The switch will have two screw connections and it should be placed on the positive side of your circuit. The switch will open and close the power going to the light fixture. However, you cannot use dimmer switches in a 12-volt DC lighting circuit.

# PLUGS

Now that you have all these nice outlets, you are probably wondering how to use them. Let's start with the devices that usually use a wall wart or power pack. Remember that wall warts and power packs are AC adapters that convert AC power into DC power to charge DC devices. Since your home is now running on DC, however, you will no longer need an adapter.

First, verify that the output of the wall wart or power supply is 12 volts DC. If it is a different voltage, then you will need one of the boost or buck converters mentioned on page 128. If the output of the wall wart or power pack is AC and not DC, then you are stuck, as this particular device will only work on an inverter.

Once you've confirmed that your power pack outputs 12 volts DC, look at the cable that goes into your device. There is usually a round jack on the cable and a corresponding socket on the device that the cable plugs into. This round jack has two parts: the round sleeve and a central pin or receptacle.

For this next step, you will need a 120-volt AC connection. Use your inverter if you have one, or go visit a friend on the grid and use one of their outlets. Plug the wall wart or power pack into the AC outlet and connect your voltmeter on a setting of at least 12 volts DC. Connect one probe to the side sleeve of the jack and the other probe to the center of the jack. It does not matter which probe makes which connection. Look at the display on the meter, and if there is *no* minus sign displayed, the red probe is connected to the positive side of the jack and the black probe is connected to the negative side. If there *is* a minus sign displayed, the red is on negative and the black on positive. Write down which is positive and which is negative and unplug.

Now, cut the wall wart from your cable so you are just left with a cable that has the jack on one end and nothing on the other end. At the bare end, strip the insulation to find two wires. Get yourself a basic AC plug at the hardware store, like the type used to repair power cords. Look for the type of plug that uses two screw terminals to connect the wires, and not the style that presses together on zip cord. There will be instructions on how to connect your wires to the plug.

Checking the voltage of the plug

Open plug

Finished plug

Connect both wires to the plug, tossing a coin to decide which goes to the narrow prong and which goes to the wide prong. We chose to connect the positive wire to the narrow prong. Plug the newly attached plug into one of your 12-volt outlets, and use your voltmeter on a setting of at least 12 volts DC to measure the voltage at the jack on the other end of your cable. Look at your notes, and verify that the polarity matches with your notes from earlier. If it doesn't, then you need to reverse the wiring on the plug because you have it wired backward. Once you have it correctly aligned, you are ready to plug the jack into your device, where it should now be running directly off your 12-volt outlet. You have now created a cable that conducts electricity directly from your house to your devices without any interruptions.

You might want to put a tag on the modified cable or spray paint the plastic part of the plug a bright color. This is to designate that while it looks just like a standard 120-volt AC plug, it is, in fact, a 12-volt DC plug. You want to prevent anyone from accidentally plugging it into a standard 120-volt AC outlet and destroying your device.

## 12-VOLT NATIVE DEVICES

You will find a lot of equipment that is designed to run off 12 volts DC. Truck stops and camping stores have many choices. Coffee makers, fans, televisions, radios, blenders, heaters, vacuum cleaners...the whole nine yards. All of these devices will have standard automotive cigarette lighter plugs, as they are designed to plug into your vehicle's cigarette lighter socket. Earlier when we discussed boost converters and buck converters, both of these devices come with cigarette lighter plugs.

If you wish, you could cut off the cigarette plug and replace it with an AC plug, using your voltmeter and the method described just above. However, once you do this, you can no longer use the device in your vehicle, as you've cut off the cigarette lighter plug. We have a better method.

At RV and camping shops, stores like Radio Shack, or on Amazon, you can purchase adapters that allow you to plug in multiple cigarette adapter plugs into one outlet. These are like the "Y" adapters used for solar panel connections that we discussed on page 110. This adapter will have one male plug and two or more female  sockets. Each female socket will have a short wire connecting it to the single male plug. Buy one with as many female sockets hanging off it as possible. We have found these with two, three, and even four sockets.

Cut the wire as close as possible to the male plug. You will end up with a bunch of female sockets on short wires. Strip back the insulation and connect the two wires inside to the same type of replacement AC plug that you used for the wall warts and power packs.

Plug the newly adapted plug into one of your 12-volt outlets. Test the polarity by setting your meter at 12 volts DC or higher and placing the red probe on the center button at the bottom of the cigarette socket and the black probe on the inside wall of the socket. Make sure you read about 12 volts without a minus sign. If you see the minus sign, then the plug is wired backward and needs to be corrected.

This is a picture of what you should have when you are done.

12 volt DC outlet to cigarette lighter socket adapter

Now you can keep your devices that have a cigarette adapter plug just as they are. Just plug the adapter you just built into your 12-volt DC outlet (do NOT use with a 120 volt AC outlet), and then plug in your device's cigarette adapter into the new adapter. You are good to go!

# DUAL CONNECTION

Perhaps the idea of keeping the grid connected to your house with the option to switch over when needed is tempting. We would like to discourage you from going this route, but if you wish to keep your grid power, you have two choices. You can duplicate your AC grid wiring with a 12-volt DC wiring plan, or you can stick with your current AC wiring, and convert it over when need be.

With either method, it is extremely important to never mix the 120-volt and 12-volt systems. Of the two methods, it is safer to wire in two complete systems if you wish to keep the grid rather than going through a conversion when the time comes.

## CONVERSION

When the grid goes down, you will need to convert your house over to 12 volts DC on all the existing wiring. Once you have made certain that the grid is physically disconnected from your house, you will need to connect 12-volt wiring to replace the grid wiring. Use the instructions starting on page 128, where we discussed how to remove the two grid cables from the main fuse/breakers, replace them with wiring to the solar-system busbars, and add a jumper between the two halves of the service panel as described earlier.

Then, you will have to replace all of your 120-volt light bulbs with 12-volt bulbs and convert all the 12-volt output wall warts and power brick supplies with plugs as described earlier. Anything needing DC voltage higher or lower than 12 volts will require either a boost or a buck converter. Some devices may need to be completely replaced with their 12-volt counterparts, like vacuum cleaners.

Make certain that any heavy load, like an air conditioner, or any 240-volt appliances, like refrigerators, electric ovens, electric dryers, washing machines, and dishwashers, are unplugged. Anything that cannot be run off 12 volts DC or through a buck or boost converter will need to be disconnected. These devices could be powered off a separate generator, or within reason, through an inverter.

When you do convert your existing AC grid wiring, you must disconnect the grid at your utility pole. Otherwise, if the grid were to come back online and you still had all your 12-volt DC devices in use, the grid power would destroy every one of them including your charge controller, and a fire would be very likely.

When switching between systems, you must take the steps to disconnect from the utility grid completely. In the next chapter, we will discuss using a transfer switch to switch between grid power and inverter power safely. This transfer switch can also be used for converting your existing grid wiring to 12-volt DC wiring, and will assure that your connection to the grid is never connected to your 12-volt power. When switched, it disconnects the grid and connects the 12-volt system.

# DUPLICATE WIRING

This approach will require twice the wiring: one set of wiring for the existing grid power and another for your DC system. Double the wires, double the switches, double the outlets, and double the fixtures.

But in return, you will have assurance of safety because you will never cross the systems. You will have double the devices in your home, 120-volt versions and 12-volt versions. The outlets and light fixtures will look identical, unless you mark or color them in some fashion to identify which are 12-volt and which are 120-volt. You don't want to take one of your 12-volt light bulbs and screw it into a fixture with 120 volts.

Hopefully, you can understand now why we recommend the cold turkey approach. If you are starting out with no pre-existing grid connection, great. Otherwise, go out to the pole, throw the disconnect switch, and call the power company to come get their meter and discontinue service. If you are uncomfortable with having no grid power, see if the power company will put an outdoor outlet out on the pole as the only grid powered outlet. You can use it with an extension cord to run tools outside, to a window air conditioner for those sweltering days, or maybe to a grid-powered outlet in a shed you want to have power in, perhaps for a freezer.

# CHAPTER 14
# PUTTING YOUR POWER TO USE
## Adjusting to DC Power

## LIGHTING

We have put hours into evaluating 12-volt LED lighting and would like to save you some time by sharing our findings.

Light bulb brilliance, for years and years, was rated in watts. By now, you know that watts really refer to the power used by a device; they do not measure brilliance. But because everyone understands how bright a 50-watt bulb or a 100-watt bulb is, watts got stuck in our heads as measurements for brightness.

Recently, lighting that is more efficient became available. By more efficient, we are saying that there are newer bulbs that give out the same brilliance as older bulbs, but require less power to do so. For example, an efficient bulb that emits the same brilliance as an older 50-watt bulb may only use 15 watts to do so. We are no longer comparing apples to apples.

Because of the huge change in lighting technology, the idea of equivalence was created. When you look at an efficient bulb, you will see two ratings: one for how much the bulb actually uses to light itself, and one for how it compares to an older bulb in brilliance. This second rating is the equivalence.

Back when we started our off-grid adventure, we found only one supplier of 12-volt LED bulbs that were efficient, and they only had two bulbs available. They offered a 50-watt equivalent bulb that pulled 3 actual watts, and a 75-watt equivalent bulb that pulled 7 actual watts. We found many other 12-volt LED bulbs that pulled many watts and were not efficient at all. Consider for a moment how these bulbs relate to amps. You know how many amps your wiring and fuse/breakers are rated for, and you know how many amp-hours of storage you have in your battery bank. You want to get as much as possible out of both. The 75-watt equivalent bulb that pulls 7 actual watts will pull 0.58 amps. Other 12-volt LED bulbs we looked at pulled as many as 4 amps!

It is now four years later and there are many more choices in efficient 12-volt LED lighting. And the prices are much better now, too. Those two bulbs we started out with cost over $30 each, and we are now able to buy more efficient bulbs for less than $10 each. We have settled on using two of the Liroyal brand bulbs: the 50-watt equivalent, and the 75-watt equivalent bulbs.

Even though these are all 12-volt LED bulbs, you will notice that they all use the standard screw in base. This base is known as an E26 base. This is a great feature because it allows us to use these 12-volt bulbs in standard fixtures. It also allows us to use them in standard table or floor lamps. You can just screw in the bulb and plug the AC-style plug into one of your 12-volt outlets.

# FANS

While air conditioning is out of the question for a solar electric system of our capacity, fans can make a major difference. We started with tabletop fans from a company called O2Cool. We found an 8-inch model and a 10-inch. These fans run on 12 volts, so it was an easy conversion from wall wart to plug to work with our 12-volt

O2Cool tabletop fan

wall outlets. They move lots of air, and using DC induction motors they are very efficient, pulling just 6 watts (0.5 amps).

Next, we installed a large 60-inch 12-volt ceiling fan, available from Sunshine Works. This fan moves almost 1,500 cubic feet of air per minute while pulling

Sunshine Works ceiling fan

only 6 watts (0.5 amps), the same amount of power as the O2Cool tabletop fan. There is a highly recommended variable-speed wall switch that also serves as a direction-reversing switch. Besides allowing the fan to run at lower speeds, this switch will also boost the voltage for higher speeds, allowing it to pull 3,925 cubic feet per minute at 13 watts.

# WATER PUMPS

We now have four pressurized water systems, one in the bathroom for the shower, one for watering the north garden, one for the kitchen's hot and cold potable water supply, and a fourth for using outside and in the mud room. These all use the same 12-volt pumps, each providing 60 PSI of pressure at 4.5 gallons per minute. The pumps are made by Flojet and sell for around $189. We learned that not all pumps are created equal when we tried a couple of less expensive pumps and found they would not hold water pressure when not in use. The water pressure would leak back through the pump, causing it to turn itself back on frequently to re-pressurize the line. Flojet is well worth the price.

The kitchen pump and the utility room systems include accumulators. An accumulator is a small tank with an internal bladder that sits between the pump and the faucet. It holds the water

Pump

pressure constant and reduces the number of times the pump switches on and off. We originally bought a complete marine system that had the pump, accumulator, and fittings all mounted on a backboard, but for the kitchen system we bought the parts and built it ourselves to save money. The pumps cost around $189 and the accumulators cost around $50. Compare this to the complete, ready-to-go marine system, which cost around $400.

Accumulator

# ELECTRONICS

Our 19-inch television came with a power brick that used 120 volts AC to supply the TV with 12 volts. All we needed to do was replace the power brick with a standard AC plug, which allowed us to plug the TV into one of our 12-volt DC outlets. The power draw is reasonable, at 24 watts (2 amps).

We chose not to have cable or satellite television service as neither were very "off-grid." We use an outdoor over-the-air antenna to keep up with the news.

We also bought a media player made by LG (model BP-530), which plays DVDs, streams Netflix, YouTube, etc., and allows us to watch movies from a USB drive. This player also uses 12 volts DC from a wall wart, so the same simple conversion that we did for the TV was done for the media player: We cut off the wall wart and replaced it with a standard polarized AC male plug

Most Wi-Fi routers use 12-volt wall warts. If possible, look at the wall wart in the box at the store before purchasing to verify it outputs 12 volts DC. We use this style router, and it gives us the Internet around the house to all our devices.

We get our actual Internet service over the Verizon cell system. Most cellular services offer Internet service over their networks, and it usually gives excellent rural service with little, if any, identification of your physical off-grid location. This is true especially if it can only locate one tower, as it typically takes three or more towers for accurate localization. We do not suggest satellite Internet service unless nothing else is available. It works almost anywhere, but it is expensive and is very sensitive to the weather. The cell router (MiFi) is powered by a USB adapter from our 12-volt outlets. The satellite receiver we looked at uses 12 volts from a wall wart.

## CHAPTER 15
# UPS AND DOWNS

### The Inverted Lifestyle: 120 Volts AC

We encourage you to power as much of your home on your solar power system's native 12 volts DC as possible, but we realize that you will probably still have a need for some 120 volt AC power in your home for one or two items that just can't easily run off low-voltage DC. So, in case we didn't convince you in the previous chapter to run your home purely on 12 volts of direct current, we now want to look at the more mainstream approach of using an inverter to convert your battery bank's 12-volt DC power into 120 volts of alternating current. You have two main considerations regarding your choice of inverter: its capacity and its waveform.

## CAPACITY

For the 1,000-watt solar array we discuss in this book, we suggest you match it to a 1,000-watt inverter. All inverters have a surge capacity, which is an amount that they can supply over and above their rating for very short times, or surges. Many devices have a high start-up load, which quickly settles back after it starts. For example, our fridge surges to 700 watts at startup, and a few seconds later drops back to running at 150 watts. This surge rating is usually limited to only 30 seconds, and often much less. We've seen some inverters with only a two-second surge capability. When you are looking for a 1,000-watt inverter, you will probably see that they are rated for a 1,500-watt surge. Verify how long they can provide that surge, as a two-second surge rating is not very useful.

A large part of your capacity decision will depend on what you plan on running through the inverter. It may help you to think of the whole discussion as if the inverter were a generator that runs off your battery bank instead of gasoline. A 1,000-watt inverter has the same capability as a 1,000-watt gasoline generator. Keep in mind that the "fuel" you have for the inverter is what you have stored in your battery bank and what your solar panels are providing. It is also important to remember that inverters are typically only 80 percent efficient, as we discussed on page 126. You are losing 20 percent of your power just in the conversion.

To determine what size inverter you need, add up the total wattage draw for everything you wish to run on the inverter at the same time. It will also give you a good idea on how feasible your needs are based on the size of the solar array and battery bank you have. For example, our home solar system can run all our lights, water pumps, fans, computers, computer equipment, and radios. We can even watch a few hours of TV in the evening. But there is no possible way that we can run an air conditioner, a washing machine, or dishwasher on our system (see Chapter 5). We do have a small microwave oven that we run on rare occasions off our inverter. During bright sunlight, we are fine, but the microwave pulls 110 amps out of the battery bank and takes a fair chunk out of our battery storage.

In our home, we have only one single item connected to our inverter, and that is our fridge.

# REFRIGERATION

For long-term food storage, we either pressure-can everything or dehydrate it. We tend to pressure-can all our meats, and dehydrate vegetables and fruits. This works great, but often we find ourselves still needing the convenience of a fridge. Even when canning and dehydrating, we need a cold place to store the items that have yet to be canned or dehydrated.

When we first went off-grid, we used a camping-style absorption refrigerator. Unlike home fridges that use a compressor, which takes a considerable amount of electrical power to run, an absorption refrigerator uses a small heat source to boil ammonia. As the ammonia boils off, the vapor passes through a number of chambers, which eventually creates a cooling effect as it condenses back into a liquid. The heat source for this type of refrigeration can be propane, electrical, kerosene, or other fuel. This type of appliance is very popular in recreational vehicles.

Absorption refrigerator

On a side note, this type of refrigeration was the first kind available for home use. Ever see those old fridges with the coil on the top? Those are absorption refrigerators. They had the unfortunate tendency to spring leaks and vent the ammonia into the house, which killed the homeowners in their sleep. Even though modern units are built better than the antique ones, be careful if you ever purchase this type of refrigerator.

While absorption refrigerators do allow for off-grid preservation of your food, they are not self-reliant. Yes, you could use your solar electric power to heat the ammonia. Most RV fridges even allow for 12-volt DC use. However, the cooling you get when using electricity to boil the ammonia is poor at best. Creating heat from electricity, whether for hot water, home heating, making coffee, or boiling ammonia, is an inefficient means of heating. With one of these fridges, you will most likely be running it off liquid petroleum gas (LP) as we originally did. And it's not very self-reliant to be dependent on a supply of LP (see page 26).

So we started looking for another solution, and what we ended up with is the most gratifying solar project we have done so far. We started with a 5.5-cubic-foot chest freezer, a fine size for the two of us. A friend is using a 7-cubic-foot model with almost as much efficiency. You'll want a new, energy-efficient model, not an old thrift store model.

There are two reasons we are using a chest freezer. First, as it is a freezer, it is designed with excellent insulation to keep the cost of retaining the cold temperatures as low as possible. The second reason is that cold air falls while warm air rises. Next time you open an upright fridge, take off your shoes and socks and stand as close to the door as you can as you open it. Feel all that cold air on your toes? That is the heavy, cold air falling out of the fridge when you opened the door. All that cold air that took your electricity to make in the first place just poured out on the ground and it is gone. A chest freezer, however, opens on the top, so when you open it, all the heavy, cold air stays inside the freezer.

You may now be thinking that our goal was refrigeration, not freezing. That is where a brewer's thermostat comes into play. Home brewers use these to control the temperature of their brewing process. These devices are electrical switches. To use them, plug the thermostat into an outlet on your inverter, and then plug the chest freezer into the thermostat. Then drop the temperature probe into the freezer. The thermostat then takes over the process of turning the chest freezer on and off, holding it at the temperature you set on its dial. To turn the freezer into the refrigerator, set the temperature to around 40°F, rather than below 32°F. Voilà! You now have a very efficient chest-style fridge! A brewer's thermostat will be available for around $50 from homebrew shops. Two good models are the Johnson Controls Freezer Temperature Controller and the Inkbird Digital Temperature Controller.

In our measurements, our fridge uses about 150 watts when running, and daily run time adds up to about 400 watt-hours. It runs every day using no external fuel and giving us never-ending cold storage for as long as the chest freezer lasts and the sun keeps shining.

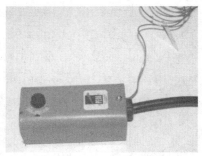

Johnson Controls Freezer Temperature Controller

# WAVEFORM

This topic gets a bit technical, so we are simply going to describe the effects and skip the engineering.

A waveform is the shape of an electrical signal. In direct current, the electrical signal doesn't change. Positive stays positive, negative stays negative, and usually the voltage level is constant. So the waveform would be described as flat.

As you know, this is not so for alternating current. AC starts at zero voltage, and in a slow rise, it reaches a maximum positive voltage. Once at the top of its curve, it starts curving back down to zero.

Once back at zero, it continues its downward curve to a negative voltage. Once it reaches the bottom of its downward curve, it starts back up to zero. Once back at zero, it has finished one cycle—zero, positive, zero, negative, zero.

We said a slow curve a minute ago, but that's a relative term. If you remember, it takes a 30th of a second to go from zero to maximum and back to zero, and another 30th to do the second half of the cycle to minimum and back to zero. The pattern that this movement creates is called a sine wave. This waveform comes from grid energy and is what all AC devices expect to receive. Unfortunately, this waveform is difficult and expensive for inverters to produce.

A much simpler waveform is one that simply turns on to maximum, reverses to a minimum 1/30th of a second later, then continually reverses from full maximum to full minimum, back and forth. This is called a square-wave. While easy and cheap to produce, a square-wave is hard on electrical devices, especially anything with a motor or a transformer. Powering the compressor in your repurposed chest freezer with a square-wave inverter would shorten its life. Some devices will refuse to run at all on a square-wave.

Inverter manufacturers came up with a compromise: the modified sine wave. They took the square-wave and rounded the edges.

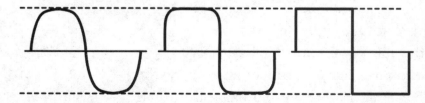

From left to right: sine wave, modified sine wave, square wave

The reason that we mention all this is to let you know that a cheap, square-wave inverter may damage your AC devices, and while cheaper than a true sine-wave inverter, a modified sine-wave inverter is only marginally better than the square-wave inverter.

We have used two brands of true sine-wave inverters, and can recommend both. Xantrex and Samlex cost between $300 and $400.

# CONNECTING AN INVERTER

If your house is already wired for 120-volt AC power, then you might wish to connect your inverter directly to your house wiring. Keep in mind a few things if you will be doing this.

First, the grid power coming into your house comes in as 240 volts, not 120 volts. It comes in a ground wire and two "hot" wires, which are each connected to the two main fuses/breakers in your circuit panel. Using your voltmeter set for AC and 200 volts or higher to measure the space between either one of the "hot" wires and the ground wire, you will get a reading of 120 volts. If you measure between the two "hot" wires, you will read 240 volts, but make sure you have your voltmeter set on a higher setting that 240 volts AC.

As we described on page 130, the breaker or fuse panel from the grid is divided into two vertical columns of breakers or fuses. We'll discuss that in greater depth now. You might think that the right side links to one of the hot wires and the left links to the other hot wire. In reality, in both columns, every other breaker is connected to one of the hot wires, and the next breaker is connected to the other hot wire, alternating back and forth. Look in your panel, and you will probably see one or more double

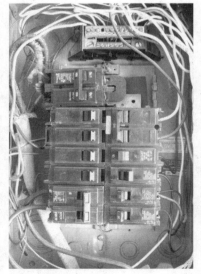

Circuit breaker box

breakers. These side-by-side double breakers connect to both hot wires, therefore controlling a 240-volt circuit and allowing them to power an electric clothes dryer or oven.

Despite the 240-volt capability of your main panel, your inverter only has a single 120-volt output even if it has two outlets. The second outlet is only for convenience. If you wish to connect your inverter directly to your house wiring, you can only connect it to one side of the fuse/breaker panel. However, because of the 240-volt nature of your panel, only half the circuits in your house would have power and the other half would essentially be dead.

The solution to this "dead circuit" problem is to tie the two halves of the circuit panel together so both sides of the panel would get 120 volts AC. Tying them together would still only give you 120 volts, so you would have to turn off all the double breakers that go into your 240-volt devices. The best way to tie the two halves together is with a professionally installed transfer switch, which will be discussed shortly.

You also need to consider if any of your house circuits are going to pull too much power for the capacity of your inverter. For example, one of your house circuits may have a window air conditioner connected, which would pull too much power. It is best to turn off all of your breakers and then turn on only those that are needed and can be run by your inverter.

When directly connecting your inverter to your house wiring, you also have to take careful consideration of any grid connection you may have. You can only have one connected at a time, never both. You must disconnect your inverter from the house wiring when you are connected to grid power. And when connecting your inverter to the house wiring, you must disconnect the grid by turning off the connection either at your utility pole or at the main breakers/fuse in your service panel.

Say that a storm has taken down your grid power and you switch over to your solar-powered inverter. If you fail to disconnect the connection to the grid, your inverter will be destroyed when the grid power is restored. Of a larger concern, if you fail to disconnect from the grid while your inverter is supplying power to your house, your inverter power will be going back out the grid wires. If a lineman is working on the problem to restore power, they will believe all the lines to be dead while they work, and your inverter power could kill them if your inverter power were back-feeding the grid lines.

We cannot emphasize this enough, especially when there is an easy solution in the form of a transfer switch. For decades, people have been buying generators and tying them into their house wiring using transfer switches. A transfer switch does two things. First, it disconnects the grid power and connects the inverter, and second, it ties the two 120-volt sides of the circuit panel together. When you want to switch back to grid power, it does the opposite. A transfer switch *must* be used to tie your inverter directly into your house wiring, and we strongly suggest you have a licensed electrician supply and install the switch for you.

## THE EASIER SOLUTION

Most inverters are equipped with two standard **ground fault interrupters**, which shut off a circuit that is going along an unintended path, such as current passing through your body. The easiest and safest way to use your inverter is simply to plug whatever you wish to run directly into the

Samlex inverter

inverter's ground fault outlet. This is how we power our modified chest freezer refrigerator. We simply use a heavy-duty extension cord and plug the fridge into the inverter. All done, mission accomplished.

# GRID TIE: PUTTING MONEY IN YOUR POCKET

Though this topic really does not fit into the scope of self-reliance or off-grid alternative power, we still want to touch on it, as it may be something you've considered. Some people with no interest in alternative power are interested in saving or even making money.

Many of the big solar re-sellers are actively pushing their products for exactly this purpose, to make solar electric power and sell it back to the grid. There are ways to connect your solar electric system, through special grid-tie inverters, to push any excess power back to the grid. Your electric meter would run backward, and the utility company would lower your monthly bill accordingly or perhaps even cut you a check if you made enough power. These grid-tie systems usually do not use a battery bank, as they are intended to reduce your electrical costs during sunlight hours rather than act as a full alternative power solution.

While this may sound good, it takes a very large system to do this, the grid-tie inverters are expensive, and every aspect of the installation has to be pre-approved by the electric utility. Every step must meet code and be installed by licensed electricians. If you are still interested in a grid-tie system, your next best step is to talk with a solar power business in your area. Consult the yellow pages.

# APPENDIX A: ZONING OUT

Neither my wife nor I are zoning or code compliance officers, nor have we ever dressed as either for Halloween. We do, however, feel that we need to touch on these topics.

If you are fortunate to live in a rural area like where we live, then you may not have to worry about code compliance. Some rural counties do not place any restrictions on unincorporated areas, but most municipalities, villages, or cities do place restrictions on your electrical power activities. Start with the zoning sections on your local and county websites. For peace of mind, a visit to the local city/village hall and the county courthouse to talk to their zoning department is also a good idea.

Zoning and code compliance rules are in place to protect you and those around you from injury, and even more so to protect future buyers of your property. You probably love your self-reliant or off-grid property and could never dream of leaving. Arlene has stated that of all the places we have lived, she could never willingly leave our current off-grid property. But never say never! There may come a reason you can't currently fathom that would cause you to consider another move, leaving you with the ordeal of selling what you have now. And not many real estate agents will be thrilled at the prospect of trying to sell your solar-powered, off-grid, self-wired, self-built home with only a wood burner for heat and a rain catchment system for water. The lack of code compliance will scare most prospective buyers away. You might be best served by running a classified in one of the homestead or prepper magazines.

If you have followed our recommendation to run your home on 12 volts direct current, you will have the advantage of only being concerned with any electrical ordinances for low-voltage wiring. Likewise, if you are plugging 120-volt alternating current devices into your inverter and not running it into any house wiring, you will not have much to worry about.

Solar PV–powered homes are covered in section 690 of the National Electrical Code (NEC), and while the NEC is not law, it is often mandated by local ordinance. You can read section 690 online at www.nfpa.org and at least familiarize yourself with what is recommended. Smaller rural homes may be ignored by building inspectors, especially when no local codes apply, but other areas may require inspection before an occupancy permit is granted. Your county or local municipality may also require that you obtain a building permit before you start to work on your system.

Even if you are under local code ordinances, you should not have any issues or problems if you keep your system "portable." If you can move it or pack it into your car, it is a portable system and not a house remodel. The 1,000-watt, four-panel, four-battery system we've described in this book may not appear to be a portable system, especially if it is tied into your house wiring. But if you would be more comfortable with a smaller and more portable system, you could install a pair of series-wired GC2 batteries in marine battery boxes that connect to an inverter and a small PWM controller, rest them on a handcart, and connect it all to couple of panels that lean against the house.

We should also touch on the topic of insurance. If your area does require permits, code compliance, or inspections and you chose to ignore any of them, your insurance company is likely to invalidate your insurance if you later suffer any sort of home loss (related or unrelated to your home power system). This is assuming you could even get insurance in the first place; our home is not insurable due to its wooden structure, wood heat, and "just because" of its off-grid nature.

We don't want to scare you away from using alternative energy; after all, when the day comes that you need to be self-reliant, your continued availability of an electrical energy source is most important. Our goal in this section is simply to make you aware of what you may have to consider as long as rule of law exists. It is cheaper to do things right the first time than have to deal with an inspector requiring you to do it over.

# APPENDIX B: EXAMPLE MIDNITE SOLAR CLASSIC CONFIGURATION

In Chapter 11, we discussed the usefulness of the charge controller. This appendix is for readers that will be using the MidNite Solar Classic charge controller manufactured by MidNite Solar. While you are of course free to use any charge controller you wish (we have used a few different ones ourselves over the years), we settled on the MidNite Solar Classic for our own use and recommend it to our readers.

This is a very powerful, customizable, and feature-rich charge controller. As such, it can be rather intimidating to a new owner. Our first controller had a single setting where we could input the battery type. Our second controller had about a dozen settings. The MidNite Solar Classic is surely more complex, but it is well worth getting to know.

To ease the process for you, we are including the configurations that we use on our controller. Each installation will be a little different, so our settings may not be optimal for you. However, they will be a place for you to start, and perhaps they will help you understand what each setting does. In particular, the battery settings will vary according to the batteries you choose to use. We'll indicate these specific settings below. As a reminder, we use the Energizer GC2 6-volt batteries in two parallel strings of two series-wired batteries each, thereby building a 12-volt battery bank.

# MODEL OPTIONS

There are three versions of the MidNite Solar Classic charge controller: the 150, the 200, and the 250. The number represents the maximum voltage each model can accept as its input from your solar panels (or wind turbine, or micro-hydro system). All three models handle roughly the same wattage, so the higher the accepted input voltage, the lower the current output will be to the battery bank. When choosing which model of the MidNite Solar Classic to purchase, consider how you are arranging your solar panel array and determine the maximum voltage that will be sent to the charge controller. For example, a solar array consisting of four typical panels wired in series and each outputting 31 volts will send 124 volts to the charge controller. Pick the lowest-voltage model of MidNite Solar Classic that can accept your solar array's voltage, as you will want the model with the highest available output current. For most installations, the MidNite Solar Classic 150 will be the best choice. To make sure you're picking the right model, MidNite Solar has a tool on its website called a "String Sizing Tool" where you can enter the specifications of your solar array and see which model of the Classic will work for you. You can always give them a call to double check.

Within each major model, there are two sub-models, called the Standard and the Lite. The main difference is that the Lite does not have an LCD control panel, and has only a limited configuration options. The Lite is primarily intended to be used in a large installation with multiple charge controllers, in which one standard MidNite Solar Classic is put in charge of one or more additional Lite models.

There are also SL models, which are for solar use only, not for wind or hydro. While the intention was to produce a less expensive model, it didn't work out this way. More Standard models were sold than SL models, so the Standard Classic model is often found at a lower price than the SL models. And if you ever plan to connect your MidNite Solar Classic to the Internet for monitoring or control, you should know that the SL model has no network capability.

# USING YOUR MIDNITE SOLAR CLASSIC

To program your MidNite Solar Classic, you'll be using the part of the controller called the MNGP, or the MidNite Graphics Panel.

MidNite Solar Classic MNGP panel

Below the screen, you'll find a panel of buttons. The main buttons you'll use are the four arrow buttons and the ENTER button. The arrows will help you move around the menu and increase or decrease various settings. The large rectangular button in the middle of the arrows is the ENTER button. After configuring a particular setting to the value you want, you must press ENTER to save, or your setting will be lost.

You will also notice two soft buttons above of the arrows, one on the left and one on the right. Any time either of these two buttons are activated, their functions will be displayed on the LCD display, directly above each button. Nothing will be displayed above inactive buttons.

Last, there are two round buttons. The one on the left is labeled STATUS and the one on the right is labeled MAIN MENU. STATUS operates the status screens, and MENU operates the programming screens. After you have your charge controller up and running, you will primarily be using the status screens. Each time you press the round STATUS button, you will move between the various status screens. Each of these screens will show the different operational parameters of your system, such as panel voltage and current, charging voltage and current, and wattage. In our opinion, the most useful status screens are the first four. If you continue pressing the status button, it will circle around back to the first screen.

Pressing the MENU button will put the charge controller into programming mode. Each menu choice has sub-menus, and pressing the ENTER button on any menu topic will move you down into that sub-menu. Pressing the MENU button will move you back up a level; repeatedly pressing MENU will eventually move you back to the home position.

## PROGRAMMING THE WHIZ BANG JR.

There is one exception to what we just told you: If you have purchased and installed the Whiz Bang Jr. monitoring accessory we recommended in Chapter 12, you will have to program its battery bank capacity from the fourth status display screen, rather than from the programming menu. (There are, however, additional Whiz Bang Jr. settings that you must program from the Main Menu.)

To set up the Whiz Bang Jr.'s battery capacity, hit STATUS until the Whiz Bang Jr. status screen is displayed. In the bottom right corner of the display panel, you will see SETUP. Press the soft key on the right-hand side, underneath the word SETUP, to enter the SETUP mode. Now you can enter your battery bank's total capacity so that the Whiz Bang Jr. will know how large your bank is. Use the up and down arrows to modify this value to match your bank. Remember that batteries in series do not add in amp-hours (see pages 73 and 80).

Now, use the arrow buttons to move over to the EFFICIENCY setting. This is where you tell the Whiz Bang Jr. how efficient your batteries are. The efficiency setting takes trial and error to find the right value, and you will never get it exactly right. Batteries are not

100 percent efficient; most are 94 to 96 percent efficient. This value will drop as the batteries age. We suggest you start with a setting of 94 percent. Press ENTER to save both of these settings.

The Whiz Bang Jr. allows your controller to tell you at a glance what your battery bank's state of charge (SOC) is; it acts like a gas gauge. It measures how many amp-hours are removed from the battery bank, and how many go back in. You can see how the efficiency setting impacts the SOC, since the amp-hours going back into the battery are affected by how efficient the batteries are at storing them. More than just the amount of amps used must be put into a battery to fully charge it. For example, if you start with fully charged batteries, and then use 10 amp-hours of their stored power, it will take 10 amp-hours plus their inefficiency to bring them back to 100 percent full charge. If you find that the SOC is showing 100 percent before the battery has stopped taking a charge (if the amps going into the battery bank have not leveled off), then the efficiency value is too high, and you should set it a bit lower. The reverse is true, too: If the charging amperage has leveled off and the SOC is not yet at 100 percent, then set the efficiency a bit higher. The efficiency setting is a value that takes trial and error to perfect, and you will never get it exactly right.

You may note that the bottom right side of the display shows MORE, identifying that the right-hand soft key will bring up some additional settings. These extra settings deal with the effect temperature has on charging the battery bank, and you should initially be able to leave these at the factory settings.

# MIDNITE SOLAR CLASSIC SETTINGS

These are the settings that we use on our solar configuration. We have not included any settings that we left at factory default.

Remember that some of these values will differ based on your specific installation; we've included notes where variability is likely.

| CHARGE MENU | | | |
|---|---|---|---|
| | SETTING | INPUT | NOTES |
| Volts | EQ | 15.6 volts | Varies per installation. Voltage for performing equalize. |
| | Absorb | 15 volts | Varies per installation. Voltage for performing absorb. |
| | Float | 13 volts | Varies per installation. Voltage for performing float. |
| ChgTime | Absorb Hours: Mins | 4:00 | Varies per installation. Amount of time needed for performing absorb stage. |
| | EQ Hours: Mins | 3:00 | Varies per installation. Amount of time needed to perform equalize stage. |
| T-Comp | Comp-MV/ Deg C/Cell | -5.0mV | Varies per installation. Charge voltage temperature compensation. |
| | EQ Comp/ D | No | Asks if temperature compensation should be performed during equalize. |
| Auto EQ | Manual | On | Turns off automatic equalization function. |
| | Re-try | 0 | If initial equalize fails, this indicates how many days the charge controller should keep trying. |
| Advanced | Ending amps | 0.0 amps | 0.0 disables the automatic absorb end based on amps. |
| | Rebulk | 12.5 volts | Varies per installation. Indicates at what voltage bulk charge should restart. |
| | Skip | 0 | Varies per installation. Indicates how many days, if any, charging should be skipped. |
| | Classic/ Shunt | | Select "Shunt" if Whiz Bang Jr. is installed; otherwise, select "Classic." |

| CHARGE MENU | | | |
|---|---|---|---|
| | SETTING | INPUT | NOTES |
| Limits | Output | 55 amps | Varies per installation. Limits the maximum current to the batteries. |
| | Input | 32 amps | Varies per installation. Limits the maximum current from the power source. |
| | Mint Comp | 13 volts | Varies per installation. Minimum voltage to use temperature compensation. |
| | Max Volts | 16.5 volts | Varies per installation. Maximum voltage to use temperature compensation. |
| | Hi Batt Temp °C | 49.0 °C | Varies per installation. Maximum battery temperature allowed to continue charging. |

| AUX MENU | | | |
|---|---|---|---|
| | SETTING | INPUT | NOTES |
| | Aux 2 | Whiz Bang Jr. | Sets up the AUX 2 input to use the Whiz Bang Jr. |
| MODE MENU | | | |
| | On/Off | On | Turns charge controller on or off. |
| | Function | Solar | Selects power source. |
| MISC MENU | | | |
| MNGP | LED mode | LED 1 | LEDs display errors, warnings, and info. |
| | Backlight | 200 | |
| | Contrast | 620 | |
| | Volume | 600 | |
| | Password | None | |
| | Time | | Set date and time to your local values. |
| | PWRSV | Off | Display back-light setting. |
| NET MENU | | | |
| | | | Connect charge controller to Internet for remote access. You will need to set these correctly to match your network. Note that the MidNite Solar Classic only has a wired Ethernet connection; it does not support a WiFi connection unless you connect it to an Access Point adapter. |

| TWEAKS MENU | | |
|---|---|---|
| **SETTING** | **INPUT** | **NOTES** |
| Offset Vbatt | 0.0 | Varies per installation. Adjust displayed battery voltage to match actual battery voltage. |
| Offset VPV | 0.0 | Varies per installation. Adjust displayed panel voltage to match actual voltage. |
| AF | On | Arc fault feature enable. |
| GF | On | Ground fault feature enable. |
| LMX | On | Pull down incoming voltage to match battery voltage. |
| ARST | On | Reboot controller at midnight. |
| Insomnia | Off | Hydro feature prevents controller from resting. |
| Nitelog | On | Log usage during the night. |
| Shade | On | Detect and display partial shading of panels. |
| TMSYNC | On | Time set locally. |
| Followme | Off | Enable multiple controllers to connect to main controller. |
| PSWD | Off | Password security. |
| WBRST | Yes | Set net AH to zero when Float is reached. |

# APPENDIX C: GRAVITY OF THE SITUATION

This appendix only applies to flooded lead acid (FLA) batteries, also known as wet cell.

Back in Chapter 10, we discussed how to determine the state of charge (SOC) of your wet cell batteries. In that discussion, we mentioned that a tool called a hydrometer could be used to measure the specific gravity of the acid in your battery. This measurement not only reveals the most exact reading on the state of charge, but it also measures the health of your batteries.

The first step is to go put on your safety glasses. Taking a measurement of your battery's acid is so dangerous that you should not even read these instructions without wearing the glasses. Got them on? Have a friend look at your face and verify you are wearing them. Positive you have them on? Okay good, continue reading.

All joking aside, you should always wear safety glasses any time you are going to remove the caps on your batteries. It is good practice to put them on before you even open your battery box.

Let's define specific gravity (SG) as a comparison of the density of battery acid to the density of water. If you had a wet cell battery, such as the GC2 batteries we use, and filled it with only water, the battery would not produce any electricity because there would be no chemical reaction between the lead plates and the water. Your specific gravity would be 1. In a typical, healthy, fully charged FLA battery, you have a mixture of 65 percent water and 35 percent sulfuric acid, measuring in with a specific gravity around 1.265 at 80°F. A higher acid content creates a higher SOC, which in turn measures a higher specific gravity.

As a battery is discharged, its acid is depleted by the chemical reaction, and as a battery is charged back up, the acid is returned to the solution. This weakening and strengthening of the acid in the solution as the battery is discharged and charged is measured by the specific gravity of the solution. As this is a direct measurement of the battery's chemistry, it is the only true measurement of the SOC; all other methods of measurement are indirect approximations.

As stated above, you can measure the SG of battery acid with a tool called a hydrometer. Not all hydrometers are created equal. Earlier we mentioned that a fully charged FLA battery measures 1.265 at 80°F. Why did we mention the temperature? Because the temperature of the acid will change the reading of the specific gravity. You have to add .004 for every 10°F above 80°F and subtract .004 for every 10°F below 80°F.

Hydrovolt hydrometer

Wouldn't it be nice if a hydrometer could do this for us? And wouldn't it be nice if it was easy to operate, didn't make a mess, and was easy to read? Let us introduce you to the Hydrovolt hydrometer, available from the Solar Panel Store for about $36.

You can measure the specific gravity at any time, at any temperature, to determine the current SOC. Simply remove the cap of the cell you wish to measure, insert the tube on the hydrometer into the acid, and squeeze then release the rubber bulb. When you squeeze the bulb, air will be pumped into the cell, so do it gently so you don't splash acid. When you release the bulb, acid will be drawn up into the hydrometer. Squeeze the bulb again to pump the acid back into the battery, and release to redraw acid. Repeat three times to make sure you have a good mix of the acid, which if left to sit tends to stratify into layers.

Now read the dial on the hydrometer to see the value at the pointer. In the picture on page 170, you can see my battery is at about 1.248. Carefully pump the acid back into the cell and withdraw the tube before releasing the bulb. After measuring every cell, pump some distilled water in and out of your hydrometer a few times to wash it. *Do not* use tap water.

This table shows the SOC based on common specific gravities:

| SOC | SG |
|---|---|
| 100 percent | 1.27 |
| 90 percent | 1.26 |
| 80 percent | 1.24 |
| 70 percent | 1.22 |
| 60 percent | 1.12 |
| 50 percent | 1.18 |
| 40 percent | 1.15 |
| 30 percent | 1.12 |
| 20 percent | 1.10 |
| 10 percent | 1.07 |

While a hydrometer and specific gravity directly measure the SOC and not how healthy your batteries are, a hydrometer is still a great tool to indirectly determine the health of your batteries.

We highly recommend keeping a battery log that records the SOC and SG of every cell in your bank. There are two things that you should watch for over time to know how healthy your batteries are (SOH):

**1** how close to 1.265 each of your cells are after a full charge,

**2** how much variation you are seeing between individual cells on the same battery.

Note that 1.265 is just a common reading for a full cell; your battery manufacturer may state a different value. Our GC2 batteries state that a full charge is 1.270.

Your log will reveal any trends over time if you have a failing cell or battery. If any one battery has a cell spread (how much variation there is between adjacent cells) greater than .05, the battery may be failing. If your battery overall can no longer reach 1.215 on the hydrometer after a full charge, it's time to replace all of the batteries in your bank, as you should never mix old and new batteries in a battery bank. Keep all batteries within a six-month age range.

Before you start taking your logbook readings, make sure that your batteries' cells are fully watered before being fully charged. If you need to add water, wait for the next full charge before taking your readings.

After they have been fully watered and then fully charged, give the batteries a rest period of 30 to 60 minutes with no load on the battery bank. This doesn't mean that you have to shut your house off for an hour, but you should limit the power draw for that period as much as possible. Then, measure each cell and record the value in your logbook.

In Chapter 10, we also discussed equalizing a battery. An equalization is intentional overcharging of the battery to help improve the battery's life span and overall health. The equalization will help break down the sulfation of the battery plates and convert more of the lead sulfate back into sulfuric acid, thereby raising the SG of the battery. Battery acid is boiled during equalization, so make sure you have the caps installed before starting or you will have acid spraying all over. The boiling action also mixes the solution, which as we mentioned tends to stratify into layers over time. Equalization also produces hydrogen gas; so again, verify that your battery box is well ventilated to the outside and that there are no flames or sparks in the area.

After performing an equalization, take new measurements of the battery or cell with the low readings and hopefully you will see improved health. You should also revisit the discussion on desulfators. Remember: *Never* add acid to a battery to try to improve its specific gravity.

# APPENDIX D:
# THE 24-VOLT OPTION

Back in Chapter 9, we started thinking about batteries as if they were buckets of electricity. The idea was that as you harvest power from your panels during daylight, anything that is not immediately used is poured into these buckets for later use. Once the sun goes down, you can take power out of these buckets to do useful things, like lighting light bulbs to chase away the darkness. In the system we discussed in this book, each recommended bucket was 1,248 watt-hours in size. Each battery was 6 volts and could hold 208 amp-hours. The bigger the buckets and the more of them you have, the more power you have on reserve.

As our system is a 12-volt system, our battery bank is also 12 volts. We connect two batteries in series to add the voltages, and then connect two of these two-battery strings in parallel to increase the amp-hour capacity. With these four batteries, we end up with 12 volts times 416 amp-hours equaling just under 5,000 watt-hours.

We also mentioned that two parallel strings is the maximum you should have. We said that you could go to three parallel strings (which would provide 7,488 watt-hours) if you really had to, but this is just asking for problems in battery charging and life expectancy. Never go above three parallel strings.

So, what can be done if you really need more stored power? One option is to use bigger batteries. There are batteries available that are larger than 208 amp-hours. Or, you could add more batteries.

Wait! Didn't we just say not to have more than two parallel strings? Yep, but how about adding to the series strings? We presently have two 6-volt batteries connected in series in each of the two parallel strings. If you double that to have four 6-volt batteries in each string, then do the math, you'll wind up with 24 volts. Do this in both of the parallel strings, and you will double your storage capacity: 24 volts times 416 amp-hours equals almost 10,000 watt-hours. Now we're talking!

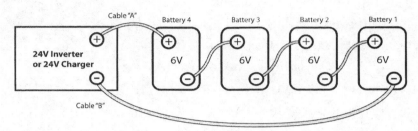

This diagram shows a single string connected for 24 volts. Adding a second identical string in parallel gives us our maximum storage.

Our charge controller can charge either a 12-volt or a 24-volt battery bank, so no changes are needed there. And if you are using a desulfator like the one we mentioned in Chapter 10, that too works on both 12-volt or 24-volt battery banks. However, if you are using an inverter to power 120-volt alternating current devices, you will need a different model than the one that was discussed in Chapter 15. You will need an inverter that converts 24 volts DC into 120 volts AC. Fortunately, the same companies we recommended in Chapter 15, Xantrex and Samlex, make both 12-volt and 24-volt true sinewave models.

Samlex 24 volt DC to 120 volt AC inverter

In Chapter 13, we tried to convince you of the benefits of running your home on 12 volts of direct current instead of the 120 volts alternating current you are probably used to. So what happens to this setup if you set up your battery bank for 24 volts instead of 12 volts? While there are many options for 12-volt devices—lights, fans, pumps, TVs, and the like—there are no 24-volt DC devices.

This is where a buck converter (page 128) comes into play. These devices lower voltage. Earlier we discussed small buck converters, designed to fit car cigarette lighter plugs and power devices that require less than 12 volts. Now we are going to introduce you to larger buck converters, which have the sole purpose of lowering

24 volt DC to 12 volt DC buck converter

the electricity flow of a 24-volt battery bank down to 12 volts for a 12-volt household. The buck converter we use is made by PYLE (PSWNV720) and available on Amazon for about $43. This converter is rated at 30 amps continuous (360 watts) and can surge to twice that for very short periods. Power at 30 amps should be more than adequate for your 12-volt house; we never come close to pulling that much power. It is a simple installation. Just connect the input side to your 24-volt battery bank, and the 12-volt output side goes to your fuse/circuit breaker panel. There are no settings to worry about; just stick it between the batteries and the loads.

Now you have yet another solution for increasing your power capacity: eight 6-volt batteries, connected in two parallel strings of four batteries each.

# APPENDIX E: RESOURCES

This appendix lists many resources mentioned throughout the book for your further exploration of the topics we have discussed. Arlene runs a blog of our off-grid experience at www.off-grid-geeks.com that you may find interesting. Alan also contributes some more technical articles in the blog's sidebar.

## PROLOGUE
### SpaceWeather
Supplies forecasts and current events of conditions in space that might impact or are impacting the Earth. They also offer a text alert service to warn of solar flares and coronal mass ejections (CME).

http://spaceweather.com

## CHAPTER 1
### U.S. Department of Energy micro-hydro information
https://energy.gov/energysaver/microhydropower-systems

### *Home Power* magazine micro-hydro information
http://www.homepower.com/articles/microhydro-power/basics/what-microhydro-power

http://www.homepower.com/articles/microhydro-power/basics/types-microhydro-systems

### U.S. Department of Energy home wind turbine information
https://energy.gov/energysaver/installing-and-maintaining-small-wind-electric-system

### *Home Power* magazine home wind turbine information
http://www.homepower.com/wind-power

### Wind Energy Foundation information
http://windenergyfoundation.org/wind-at-work/wind-consumers/wind-power-your-home

**U.S. Department of Energy wind map**
http://apps2.eere.energy.gov/wind/windexchange/wind_maps.asp

## CHAPTER 2
**Gasoline to LP conversion kits for generators**
http://www.propane-generators.com/basic_info.php

http://www.uscarburetion.com

http://www.instructables.com/id/Converting-a-generator-to-run-on-propane

**Wood Gas**
http://www.driveonwood.com

## CHAPTER 4
**Watt meter**
Kill A Watt meter from P3 Company, available on Amazon.com for under $20

## CHAPTER 5
**Lehman's Hardware store for the non-electric lifestyle**
4779 Kidron Road, Dalton, OH 44618
https://www.lehmans.com

**Solar-powered oven**
https://www.sunoven.com

## CHAPTER 6
**K-Tor human-powered generator**
https://www.k-tor.com

**Bicycle-powered generator**
http://pedalpowergenerator.com

**Review of human-powered solutions**
http://www.alternative-energy-news.info/technology/human-powered

*Low Tech Magazine*
http://www.lowtechmagazine.com

*No Tech Magazine*
http://www.notechmagazine.com

## CHAPTER 8
**Sun-angle calculator**
http://www.solarelectricityhandbook.com/solar-angle-calculator.html

**Harbor Freight solar panels**

http://www.harborfreight.com/45-watt-solar-panel-kit-10-pc-kit-68751.html

**Grape Solar**

http://www.grapesolar.com

## CHAPTER 9
### Battery vendors

Trojan (www.trojanbattery.com) and Rolls (rollsbattery.com) are excellent choices for batteries.

Fullriver AGM Batteries are also worth looking at, from RES Supply (ressupply.com). The most economical we have found is the Energizer GC2 available from Sam's Club (samsclub.com).

## CHAPTER 10
### Water distillers

http://primewater.us

### MidNite Solar charge controller and energy monitor (Whiz Bang Jr.)

http://www.midnitesolar.com

### Flooded wet cell battery desulfator

http://batterylifesaver.com

## CHAPTER 11
### MidNite Solar charge controller

http://www.midnitesolar.com

### Online monitoring service for MidNite Solar charge controllers

https://mymidnite2.com

## CHAPTER 14
### O2 Cool fan

http://www.o2-cool.com/fd10018?page_id=1

### Ceiling fan

http://www.sunshineworks.com/solar-fans-vari-cyclone-ceiling-fans.htm

## CHAPTER 15
### Inverters

http://www.samlexamerica.com

http://www.xantrex.com

## APPENDIX C
### Hydrovolt hydrometer

http://www.solarpanelstore.com

# INDEX

# ABOUT THE AUTHORS

**Alan and Arlene Fiebig** (aka Off Grid Geeks) leaped off the grid in 2012. Their goal in this pre-emptive bug out was to become as self-reliant as possible while still maintaining their high-tech lifestyle and work-from-home jobs. While Alan's 40-plus years of expertise in all things electronic and Arlene's degree in mechanical engineering proved to be beneficial in their new lifestyle, their ability to think outside the box allows them to come up with inexpensive alternative solutions for costly projects. This husband-and-wife team lectures and consults on a wide range of off-grid and self-reliance topics and are appreciated for their ability to explain technical subjects in an easy-to-understand manner. You can learn more about going off the grid at their website, http://off-grid-geeks.com.